Video Camera Technology

For a complete listing of the *Artech House Audiovisual Library,*
turn to the back of this book.

Video Camera Technology

Arch C. Luther

Artech House
Boston • London

Library of Congress Cataloging-in-Publication Data
Luther, Arch C..
    Video camera technology / Arch C. Luther.
        p.   cm. — (Artech House audiovisual library)
    Includes bibliographical references and index.
    ISBN 0-89006-556-X (alk. paper)
        1. Television cameras.   I. Camcorders.   I. Title.   II. Series.
    TR882.L88   1998
    778.59—dc21   98-10936
                                                       CIP

British Library Cataloguing in Publication Data
Luther, Arch C.
    Video camera technology
    1.Video tape recorders   2. Video recording—Technique
    I. Title
    778.5'99
    ISBN 089006556X

Cover design by Jennifer L. Stuart
Cover illustration by Matthew Neutra

© 1998 ARTECH HOUSE, INC.
685 Canton Street
Norwood, MA 02062

International Standard Book Number: 0-89006-556-X
Library of Congress Catalog Card Number: 98-10936

10 9 8 7 6 5 4 3 2 1

Contents

Preface

Many people who watch video on television, the Internet, CD-ROM, Digital Versatile Disc (DVD), and so on, never give thought to the devices that capture video—video cameras. But they represent a highly developed technology, with a history going back more than 50 years.

In the early days of television, a video camera was a large device that cost upward of 50,000 1950 dollars, required two or three skilled technicians to operate, and used pickup tubes costing $1,000 that lasted only a few hundred hours. It produced monochrome pictures. Only large broadcast stations or networks could afford to have them.

Today, a video camera can be bought for less than $500 that is small enough to be hand held, requires no technical skill to operate, makes pictures in color, and even contains a video recorder—a device that didn't even exist in the 1950s. Millions of consumers have their own video cameras and they are available for all levels of professional and semi-professional use in many fields.

The story of how video cameras developed and improved over the last 40 years is exciting indeed and reads much like the more recent story of computer development. Orders of magnitude of improvement have occurred, resulting in almost complete product obsolescence every 10 years or so. The early video cameras, even some less than 20 years old, exist today only is museums. A complete book could be written about this history.

But history is not the subject of this book—it is about today's technology. Little space is given to the history of video cameras because current camera technology is exciting and interesting enough in itself. The current and emerging technology needs to be understood by purchasers, operators, and developers of video cameras.

This book is an exposition of video camera technology at all levels from consumer products to professional broadcast and production cameras. It is written for readers who have some technical background and are interested in the technical aspects of camera design or operation. An electrical engineering or computer science background is expected of the reader. The presentation is relatively non-mathematical, although equations are used where they can enhance understanding and use of the material.

The most recent advances in all forms of electronic technology involve using digital techniques for things that were previously done with analog circuits. Video cameras and television equipment in general are some of the forerunners of this trend. Digital technology offers higher performance, lower cost, and greater reliability in all forms of video

xvi Video Camera Technology

equipment. This trend has already reached the point where nearly all new cameras are digital and, therefore, this book focuses on digital techniques for cameras. Analog technology that is rapidly going out of use is not covered, in favor of the newer digital techniques.

Most modern cameras are *camcorders*—they also contain a video recorder. Recording is another major technology that could deserve its own book. It is covered here in one chapter that relates to camcorder applications (Chapter 8).

Video camera technology has become an important part of the information infrastructure that provides video communication around the world.

ACKNOWLEDGMENTS

This book would not have been possible without the assistance of many members of the camera industry who provided information, advice, and comments. I particularly want to mention Tom Leacock and Stan Basara of Panasonic, Arnold Taylor and Craig Risebury of Cintel, and John Smiley. But especially, I want to mention Larry Thorpe of Sony who spent hours with me throughout the project. And, lastly, I am constantly amazed at the information on this (or any other) subject that is available on the World Wide Web. Because of the importance of that source, there is a special section in the Bibliography for Web uniform resource locators (URLs) related to video cameras.

1

Introduction

History shows that humans have always been interested in the reproduction of natural images. From prehistoric cave drawings to medieval paintings, to chemical photography, steady progress has been made in both the technology and the art of imaging. The most recent stage of that progression is *electronic* imaging—the capturing, processing, and displaying of images in the form of electronic signals—*video*. This book is about devices for capturing electronic images—*video cameras*.

1.1 VIDEO SYSTEMS

Electronic image reproduction always requires a *system* consisting of several units. In the simplest case, a camera is connected directly to a display, but most systems involve other functions as shown in Figure 1.1. A signal source, usually a video camera, converts a natural scene into electronic signals that are processed, stored, or transmitted according to the system requirements. Finally, a display device converts the signals to a form that can

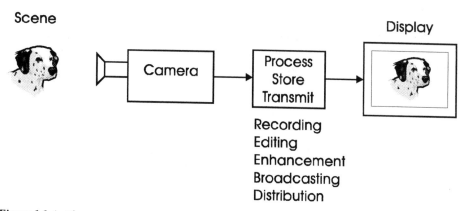

Figure 1.1 *A video system.*

Table 1.1
Camera Requirements for Different Applications

Attribute	Home	Semiprofessional	Production	ENG
Picture quality	VCR-level	Standard TV	Highest quality	Standard TV
Optical features	Zoom	Zoom, filters	Zoom, filters	Zoom, filters
Automatics	Exposure, focus, white balance	Exposure, focus, white balance, optional manual	Exposure, white balance, optional manual	Exposure, focus, white balance
Preferred configuration	Camcorder	Camcorder	Camcorder or Camera	Camcorder
Mounting	Hand-held	Hand-held or tripod	Tripod or pedestal	Hand-held
Size/weight	2–6 lb	2–10 lb	15–100 lb	2–10 lb
Synchronization	None	None	Yes	None

be viewed by the end user. A home video system may consist of simply a camcorder (combining the functions of camera and recorder) and a display device, such as a television receiver or monitor. However, broadcast stations have much more complex systems consisting of multiple cameras, recorders, switchers, editing controllers, effects units, signal processors, and so on (see Chapter 6). The requirements for a camera are very much a function of the system in which the camera will be used; a few examples are shown in Table 1.1.

1.2 USES OF VIDEO CAMERAS

Cameras have many uses, which may be segregated according to whether a motion picture is broadcast to many viewers (television), a motion picture is distributed for special purposes (non-TV), or still pictures are captured.

1.2.1 Television

Television broadcasts motion video programs to many viewers in a specified geographical area, which may be a single city or metropolitan area with terrestrial broadcasting or may include the entire world with satellite broadcasting. Essentially all TV signals originate from cameras. Early TV systems were unable to store signals, so cameras signals were delivered "live" in real time to display devices by wire connection or broadcasting. Today, with videotape storage capability, the preferred mode for most applications is to capture signals onto tape and, at a later time, assemble them into programs for replay or distribution. This is the *production-postproduction* style of program creation. Live TV is now used only in applications where events must be captured and shown as they happen, principally in sports or news. Even in these cases, backup recordings are usually made so that programming can be enhanced by effects such as slow-motion playback or instant replays. A recording also provides the capability to replay the live-produced program at a later time.

Because TV signals are produced in many locations and are broadcast to millions of viewers around the world, the matter of standardizing is extremely important. It would be ideal if there were only a single standard worldwide for television, but that has so far not been achieved. At present, there are three basic analog TV standards in the world, with some minor differences between countries; these standards are referred to as *standard-definition television* (SDTV). The worldwide standards issue is being faced anew with the development of *digital television* (DTV) and *high-definition television* (HDTV), but it is still too early to tell how many versions of that will be adopted around the world.

1.2.2 Non-TV

When motion video is created for purposes other than broadcasting to millions of viewers, it is called *non-TV*. Applications for this are videoconferencing, surveillance, laboratory research, remote observation, medical imaging, and others. In most of these applications, the environment of distribution is small and controlled, so the matter of standardizing all systems becomes less important. Because the systems are often specialized and have their own unique standards, camera requirements vary for this class of systems.

1.2.3 Still Pictures

Although most video cameras are used to produce still pictures at one time or another, a new class of electronic camera is now available that has the specific objective of taking still pictures. These *electronic photography* cameras compete with film photography. Electronic still cameras are also used in medical imaging, and document storage. Still cameras pose their own special problems, which are discussed in Chapter 11.

1.3 COMPONENTS OF A VIDEO CAMERA

A video camera views a natural scene and converts what it sees into an electrical video signal, as shown in Figure 1.2. This is accomplished by focusing an optical image of the scene onto the sensitive surface of an *image sensor*, which performs the conversion to an electrical signal by a *scanning* process (see Section 1.6). Once the signal has been delivered by the image sensors, additional electronic processes may enhance or modify the signals. Then, they are displayed or stored for later use, which may consist simply of display, but multiple signals often are assembled into a *program*.

The output from an image sensor is an analog signal; many cameras perform all their processing steps in the analog domain and deliver an analog signal output. However, there are major advantages to converting the analog signal to a digital format as early as

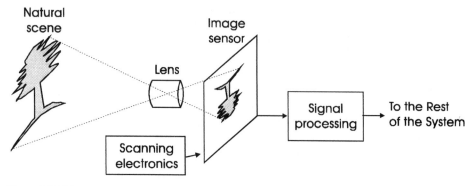

Figure 1.2 *Video camera components.*

possible in the signal path. This approach is used in most new camera designs and is explained in Section 1.5.

Optical systems are covered in Chapter 2, image sensors are the subject of Chapter 3, and signal processing is discussed in Chapter 4.

1.4 CHARACTERISTICS OF HUMAN VISION

The final judge of the pictures from a video system is a human viewer, who watches the results on a display. The design of a system and its cameras depends critically on choices and compromises that are based on the properties of human vision. With an understanding of vision, we can decide what is important or unimportant in picture reproduction.

1.4.1 Acuity

The ability of human vision to perceive fine detail is called *acuity*. A person's acuity is expressed as the angle subtended by the smallest object he or she can discern. For gray scale objects, this is typically about 1′ (minute of arc), although there is considerable variation among individuals. (Acuity is poorer for colors—see Section 1.4.3.) In a video system, acuity corresponds to *resolution* performance—the ability of the system to reproduce fine detail (see Section 1.6.1). Obviously, it is not necessary for a system to reproduce more detail than a viewer will be able to see when viewing displays from the correct distance. Because of acuity being specified as an angle, the amount of detail one can see on a screen depends on the distance between viewer and screen. This is quantified as the *viewing ratio*—the ratio of the viewing distance to the picture height. Table 1.2 shows viewing ratios for different video systems. These figures are typical values; viewers may actually sit closer or farther from the screen, which results in variation of perceived picture quality.

Determining the necessary resolution performance of a system requires first specifying a design value for the viewing ratio. Then the resolution is calculated in terms of the

Table 1.2
Typical Viewing Ratios and Resolution

System	Minimum Viewing Ratio	Resolution TVL
NTSC television	7	320
HDTV	3	1,000
Computer display	1	800

maximum number of repeated objects a viewer can distinguish in a distance equal to the picture height. Some resolution numbers are shown in Table 1.2; they are discussed in Section 1.6.3.

1.4.2 Brightness

Electromagnetic radiation in the wavelength range from 400 to 700 nm (nanometers—10^{-9} m) is what we know as *visible light*. Most natural light contains energy at several different wavelengths. For example, "white" light contains an approximately uniform distribution of wavelengths, but colored light will have a predominance of one or a few wavelengths. A human observer perceives the intensity (energy level) of light as the sensation called *brightness*. However, the perceived brightness varies depending on the color of the light. This is quantified by a curve of the brightness sensation versus wavelength, called the *luminosity curve* (see Figure 1.3). A monochrome video camera must have a spectral response that matches the luminosity curve. Such a camera generates a *luminance* signal.

The eye can detect several hundred levels of intensity within a scene, which is referred to as *gray scale* response. The range of intensity observable in a single scene is more than 1000:1 but this is seen as only a few hundred different levels. It is well known that the eye adapts to average scene brightness over an extremely wide range, as much as 10^{10} to 1. Video cameras must also deal with this brightness range and provide gray scale reproduction pleasing to the eye. This is discussed further in Section 2.1.1.

1.4.3 Color Vision

Colors are perceived because, in addition to its gray scale sensors (rods), the human eye contains three types of color receptors (cones) that have different spectral responses, and the brain performs signal processing to interpret colors from the data received. This process is modeled in imaging systems by using three electronic channels to carry data from three separate color sensors. Such systems are called *trichromatic* and they have the objective of producing a color reproduction that a viewer will perceive as natural.

Colors theoretically can be reproduced by appropriately combining signals from any three channels that have different spectral responses. However, there are many practical

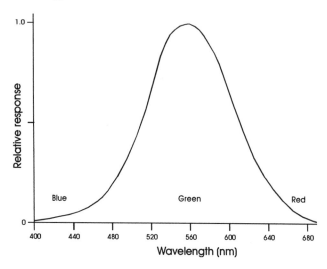

Figure 1.3 *The luminosity curve.*

considerations that narrow down the choice of channel properties and require that they be standardized for specific purposes. For example, in color printing, colors are reproduced by placing colored inks on white paper, a process known as *subtractive color* because the inks subtract colors from the light reflected by the white paper. Similarly, in color TV, colors are reproduced by combining three colored light sources—this is *additive color* because the colored lights add to produce the reproduction. Additive and subtractive color systems are discussed further in Section 1.7. The specific colors used for the three channels are called *primary colors*.

Another important property of human color vision is that the acuity of color vision is significantly poorer than for gray scale vision. This is a result of the way the eye and brain process colors, and it is exploited by most color video systems. To do that, signals must be generated that do not contain any gray scale information—these are *color-difference signals* and they are produced by subtracting the luminance signal from the color signals. Color-difference signals describe only color, without luminance—they make up the *chrominance* components of the video.

1.4.4   Motion and Flicker

It is fortunate for imaging systems that the human eye does not respond instantaneously to a scene. There is a *persistence of vision* effect caused by the properties of the eye and brain. Because the perception of a scene develops slowly in the brain and also changes slowly when motion occurs, we are able to see a sequence of discrete images (frames) as a smoothly moving picture. The key parameter of motion reproduction is the frame rate, which must be sufficiently high for us to see smooth motion. The threshold for this is in the range of 24 to 30 frames/s.

A completely separate effect is the perception of *flicker*, a disturbing phenomena caused by flashing lights. It also occurs in the display of frame sequences. The threshold of flicker depends on image brightness and viewing conditions, but it is significantly higher than the threshold of smooth motion—in the range from 48 to 75 frames/s.

1.5 DIGITAL TECHNOLOGY

Although natural images are completely analog in that brightness, color, and position are all continuous functions of time; most electronic reproduction systems process one or more of these properties as discrete values. For example, reproducing smooth motion as a sequence of frames is discreteness in time. A digital video system converts *all* signal properties into discrete values.

1.5.1 Advantages and Disadvantages of Digital

The advantages of digital technology in video systems now so outweigh the disadvantages that nearly all new systems are digital and existing analog systems are being replaced. This even extends into the home, where audio systems are already digital (the compact disc), and digital video equipment is being introduced. The advantages of digital video are

- The precision of digitally representing analog quantities is established at the point of conversion from analog to digital. Once established, the digital precision can be maintained (if desired) through an extended system by the use of digital *error protection*. This is a perfect reproduction system, which is said to be *transparent*.
- Digital signals can be stored in memory devices. Many processes are facilitated by storage, such as image enhancement based on storing adjacent data from the image, or conversion between different signal standards.
- Complex digital functions are produced inexpensively by designing them into integrated circuits. This allows extremely sophisticated signal processing techniques to be achieved easily and economically.
- One complex process that can be done is *video data compression*, which has the result that digital video transmission may be done with less bandwidth than equivalent analog video.
- Digital systems are less expensive than equivalent analog systems and the cost will continue to fall as integrated circuit technology becomes cheaper.

The principal disadvantage of digital technology is that digital circuits are inherently more complex than analog circuits. This originally made digital technology more expensive than analog technology, but that has been overcome by the use of integrated circuits (ICs), which are more effective for digital techniques than for analog techniques. Of course, that is the same reason digital systems are now often cheaper than equivalent analog systems.

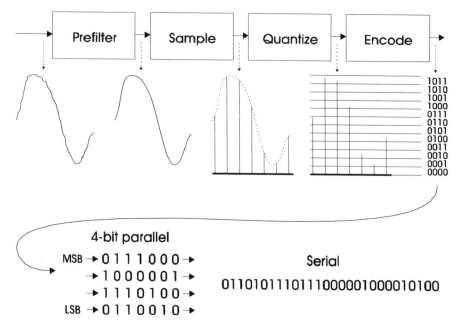

Figure 1.4 *Block diagram of an ADC.*

1.5.2 Analog-to-Digital Conversion

Most signals exist in nature as analog quantities and they must be converted to a digital format upon entering a digital system. This is analog-to-digital conversion (ADC), sometimes called *digitizing* and, as explained in the previous section, it is where the limits of system performance are determined. Every video camera that employs digital processing contains one or more ADCs. ADC is a complex process, the details of which are beyond the scope of this book [1], but there are some key parameters that will be discussed here.

ADC of an analog waveform involves three steps—sampling, quantizing, and encoding. Sampling and quantizing can theoretically be done in either order. They convert the analog signal into a stream of digital *samples*, each being represented by a specified number of bits. This is shown in Figure 1.4.

1.5.2.1 Sampling

The process of digitizing a waveform along the time scale is *sampling*. The instantaneous value of the waveform is read at equally spaced points to create a stream of *samples*. If sampling is done first (before quantizing), the sample values are still analog quantities. If quantizing is done first and then sampling, the samples come out as full digital values.

The most important parameter of sampling is the *sampling rate*, which determines the highest frequency of the input waveform that can be correctly digitized. This is according to the *Nyquist criterion*, which states that the sampling frequency must be at least twice the highest frequency component of the signal being sampled. From the signal frequency point of view, the Nyquist criterion says that signal frequency should not be greater than one-half the sampling frequency. That is called the Nyquist limit. If it is exceeded, spurious signals are generated, called *aliasing* components; this is distortion that cannot be removed later. Because of this, most applications of ADC employ a *prefilter* to make sure frequency components above the Nyquist limit are removed before sampling. The sampling frequency determines the maximum horizontal resolution of a video signal.

1.5.2.2 Quantizing

The other step of ADC is *quantizing*, which converts the analog amplitude scale into a scale of discrete levels that can be represented by a fixed number of bits. For example, reducing a signal to 256 levels corresponds to 8 bits per sample (bps). The bps parameter is the second important ADC parameter after sampling frequency; multiplying those two numbers gives the *data rate* of the ADC.

The principal artifact of quantizing is *quantizing error*, sometimes called quantizing *noise*. By itself, it is not really noise in the sense of randomness, rather, it is a signal-dependent distortion that produces a visual effect called *contouring*. This gets its name because the effect in a picture looks like a contour map. Because contouring is signal-related, it is easily visible even at low levels; most ADCs deliberately add some random noise to break the contouring patterns—this is called *dithering* [2]. With dithering, quantizing errors do look like noise.

The bottom line on quantizing is that it determines the system's maximum signal-to-noise ratio (SNR). This can be calculated from the number of bits per sample N by

$$SNR \ (dB) = 6.02 \ N + 10.8 \tag{1.1}$$

Thus, 10-bit quantizing has a maximum SNR of 71 dB.

1.5.2.3 Encoding

The relationship between the quantizing level of a sample and the actual bit pattern that is output from the ADC is called its *encoding*. There are many ways to do this, from the simple binary encoding shown in Figure 1.4 to very sophisticated processes involving look-up tables or other digital processing devices. Of course, one receiving a video bit stream must know the type of encoding used—it is part of the standardization of bit streams.

1.5.2.3 Pixels

When a sequence of samples represents a video signal, each sample can be considered to be a single point of the reproduced picture—a *picture element* or *pixel*. Each pixel has its own brightness and color values independent of adjacent pixels. This concept is also used in analog systems by viewing each half-cycle of the highest video frequency as a pixel. The number of pixels in a digital image will determine the amount of fine detail that can be reproduced by the image (see Section 1.6.2).

1.5.3 Digital Processing

An important advantage of digital video is that pixel values can be processed in various ways to perform image enhancement or manipulation. Processing of pixel values requires arithmetic or logic operations that are performed in hardware logic or on a computer. The only limitation on what can be done is speed—pixel rates are very high and complex operations on pixels require extremely high-speed processing. However, devices and software are available for many such operations.

1.6 SCANNING

An optical image under steady illumination is what computer people call a *parallel* process—all points of the picture are simultaneously reflecting energy into the lens. However, an electrical video signal is *serial*—it can convey only one value at a time. The parallel-to-serial conversion process is called *scanning*. This operates in exactly the same way one reads a page of English text. The points of the image are sensed sequentially, beginning (usually) at the upper left corner and progressing horizontally across the image to the right side. The sensing then snaps back to the left, moves down a line, and again scans horizontally. This repeats until the bottom of the image is reached, where the sensing jumps back to the upper left to begin another scan. A complete scan of the image is called a *frame*, which is made up of a horizontal scanning motion and a much slower vertical scanning motion.

1.6.1 Progressive Scanning

Figure 1.5 shows a scanning process and illustrates some key features. The scanning pattern, shown in Figure 1.5(a) is called a *raster*. It is a pattern of approximately-horizontal lines created by simultaneous horizontal and vertical motions. To produce a stable repeating pattern, the horizontal and vertical scanning frequencies have an exact integral relationship. This is known as *progressive scanning*. The number of lines is equal to ratio of horizontal to vertical frequency and is called the *line number* for the system. Scanning systems are referred to by their line number and their vertical scanning frequency; i.e., 525–30. For several reasons, it is convenient for the scanning motion to have a uniform velocity, which implies sawtooth-shaped scanning signals, as shown in Figure 1.5(b). The vertical-scanning frequency is also the *frame rate*, which is chosen to deliver satisfactory reproduction of motion (see Section 1.4.4). Since the scanning velocity and thus, the video bandwidth or bit rate, is proportional to the vertical-scanning frequency multiplied by the line number, it is not desirable to make vertical scanning any faster than necessary for reproduction of smooth motion and elimination of flicker.

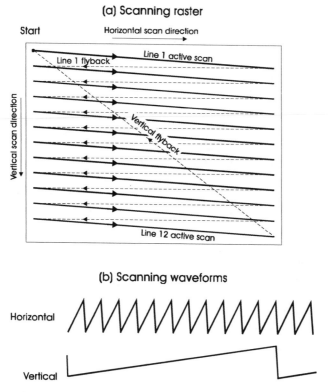

Figure 1.5 *Progressive scanning.*

Figure 1.5(b) shows that the scanning does not instantaneously return to the left at the end of each line. A finite time, called the *flyback* time, is required, and a portion of the total scan period called the *blanking interval* is allowed for it. The rest of the scan period is called the *active scan time*. These parameters apply both horizontally and vertically.

1.6.2 Aspect Ratio

Another important scanning parameter is the *aspect ratio*, which is the width-to-height ratio of the raster. Standard TV has an aspect ratio of 4:3, HDTV is 16:9, and movie theaters since 1953 have used aspect ratios in the range of 1.85:1 to 2.35:1 (CinemaScope). The attraction of the wide-screen movie configurations prompted the developers of HDTV to also plan for wider screens. Figure 1.6 shows a comparison of aspect ratios where each is shown with equal screen area. It is curious that the common area (shaded) has approximately a 16:9 aspect ratio, as also does the maximum area enclosing all the screens. This has been used as an argument for choosing 16:9 for HDTV. Another bit of magic is that 16:9 is the square of 4:3.

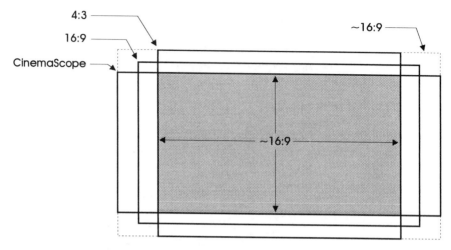

Figure 1.6 *Aspect ratios shown with equal areas.*

1.6.3 Resolution

As explained in Section 1.4.1, the viewer's acuity and viewing distance determine the desired resolution specification for a system. In analog systems, resolution is specified by counting the maximum number of black and white lines the viewer can perceive in a distance equal to the picture height. This number is the *TV line* (TVL) number specification. Assuming the 1′ of arc value for acuity, the desired TVL number can be calculated for a typical viewer as

$$TVL = 3{,}440 \: / \: VR \tag{1.2}$$

where *VR* is the viewing ratio.

The system line number must be somewhat higher than the desired resolution. This is because the line pattern performs a sampling process in the vertical direction, which limits the resolution as explained in Section 1.5.2. In analog systems, it has been customary to speak of the ratio between apparent vertical resolution and active lines. This is the *Kell factor* and, in analog systems, it has a value of approximately 0.7.

Horizontal resolution *HR* in analog systems is limited by the video bandwidth, according to (1.3)

$$HR_{TVL} = 2 \: VB \: HA \: / \: f_H \: AR \tag{1.3}$$

where   *VB* is the video bandwidth in Hz;
HA is the horizontal active-scan fraction, typically 0.82 for standard TV;
f_H is the horizontal scanning frequency in Hz;
AR is the aspect ratio.

Equation (1.3) can also be invoked in reverse by saying that a specified resolution calls for a certain minimal bandwidth.

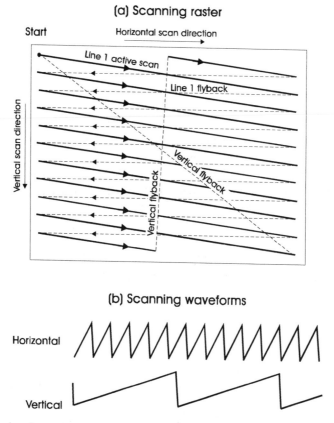

(a) Scanning raster

(b) Scanning waveforms

Figure 1.7 Interlaced scanning.

1.6.4 Interlaced Scanning

Existing TV systems use *interlaced scanning*, a technique that doubles the horizontal resolution for a given bandwidth compared to that calculated by (1.3). Half of the total lines are scanned in one vertical scan and the other half of the lines are scanned by a second vertical scan. That is accomplished by making the line number be an odd integer and scanning horizontally at one-half the vertical scanning frequency multiplied by the line number. This relationship causes interlacing of odd and even scan lines as shown in Figure 1.7. Since the vertical-scanning frequency is twice the frame rate, large-area flicker is reduced for a given frame rate. The lower horizontal-scanning frequency also lowers the cost of displays.

Interlacing has worked well for television systems that have large viewing ratios and limited resolution, but its artifacts show clearly when it is used with computer-generated images or with higher-resolution natural images that are viewed at close range. The worst problem is *interline flicker*, which is a severe flickering of horizontal or near-horizontal edges where the information on adjacent lines is not the same. On computer screens, this is evident on text characters or sharp lines. Severe filtering must be used to successfully display computer-generated images or text on TV screens; this limits the maximum size of readable text on TV screens to about 40 characters per line. Interline flicker on computer-generated information is so severe that interlacing is almost never used in computer displays.

With the advent of DTV systems, the use of interlaced scanning has become controversial. However, DTV allows the scan parameters for display to be different from the transmitted scan parameters, so that interlacing of receiver displays can be optional. Even if the transmitted signal is interlaced, a high-quality digital receiver can use a progressively scanned display which, if done properly, will eliminate all interlace artifacts. Interlaced displays will still probably be used in low-priced digital receivers but deluxe receivers can use progressive scanning. The controversy is about whether interlaced scanning should ever be the transmitted format. That has been left optional in the standards that have been adopted so far.

Interlacing has special considerations with regard to charge-coupled device (CCD) imagers (see Section 3.4.4.2; also Section 13.2.3 for HDTV interlacing).

1.6.5 Synchronization

Scanning implies precise synchronization between the video signal and the horizontal and vertical scanning used by a display. Otherwise, the display of the signal will be scrambled. In an analog TV system, scanning of cameras and displays all operate at the same rates and the signal standards provide pulses that are easily separated to control the scanning. In digital systems, however, synchronization takes the form of unique code words in the data stream that can be detected as the data flows in real time. Recognition of a sync word implies that the data immediately following has a special meaning—usually it is some form of header structure that defines the following data type and its structure. This may occur in several levels, for example, in a video stream, one sync word identifies the start of a frame and different sync words identify data blocks within that frame. Note that compressed video data almost always contains a frame structure but the lower-level structure may be very different from the line structure found in analog video.

1.7 COLOR REPRODUCTION

Human vision is sensitive to the spectral content of light, giving the property we know as color. Although an exact physical reproduction of colored light would require a wavelength-for-wavelength matching of the spectrum of the light, that is not necessary to produce a reproduction to satisfy the a human observer. Only three attributes need be matched, and there are many systems of three parameters that can be used. This section examines light spectra and systems of matching colors.

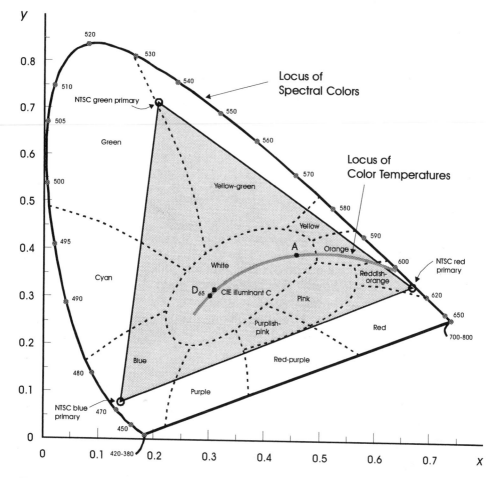

Figure 1.8 *The CIE chromaticity diagram.*

1.7.1 The CIE Standard Observer

In 1931, as the result of extensive research, the International Commission on Illumination (called CIE, for its French name, *Commission Internationale de l'Eclairage*) defined standards for color matching functions and a diagram that are still in use today. A set of imaginary nonphysical primary colors, **X**, **Y**, and **Z** were chosen that could match all the colors in the spectrum without using negative values [3]. Primaries **X** and **Z** were chosen to have no luminance, meaning that the value Y of primary **Y** carries all the luminance of a color being matched. The other two values, X and Z, thus represent only the chromaticity of the color. Values x and y are formed by normalizing X and Y against the total of X, Y, and Z. Plotting y versus x gives the *CIE chromaticity diagram*, shown in Figure 1.8.

The parameters x and y are called the *chromaticity coordinates* and can define all possible colors.

Spectral colors form a horseshoe-shaped curve on the chromaticity diagram as shown in the figure. All real colors are within the horseshoe, with "white" colors in the lower center of the curve as shown. All colors are represented by points in this space; primary colors are simply three points defining a triangle, the area of which represents the *gamut* (range) of colors that can be reproduced by those primaries.

1.7.1.1 Hue, Saturation, and Brightness

Any three independent parameters can be used to specify colors. One system that is convenient because it relates to easily observed properties is the *hue-saturation-brightness* (HSB) system.

Hue refers to the dominant color or wavelength of the sample. On the chromaticity diagram, it is quantified by passing a line from white through the color point out to the spectral curve. The dominant wavelength is where the line intersects the spectral curve. More subjectively, it is the answer to the question: "What color is it?" Words such as red, yellow, orange, pink, and so on are statements of hue.

Saturation is the intensity or depth of the color. On the diagram, it is indicated by how far the color point is from white toward the spectral line. Low saturation colors are "washed out," while high saturation colors are "intense."

Brightness is the same as the luminance parameter defined in Section 1.4.2. It is a subjective evaluation of the total energy in the color. The chromaticity diagram does not show brightness.

1.7.2 Primary Colors

According to the trichromatic theory, color imaging requires three sensing channels having different spectral responses. The optimal spectral responses depend on the colors that will be used for reproduction, these are the *primary colors*, which may be additive or subtractive (see Section 1.4.3). The additive primaries, red, green, and blue (RGB) give a large reproductive gamut as shown by the shaded triangle in Figure 1.8 and are the best choice for video systems. Specific primary colors are defined by their chromaticity coordinates x and y. The primaries shown in Figure 1.7 are those standardized in the United States by the NTSC. Display devices must be able to reproduce the primary colors and this is a major criterion for the choice of primaries.

Figure 1.9 shows the RGB additive primaries with their *complements*, which are the colors produced by combining two primaries at a time. The complements of the additive primaries are the subtractive primaries. A given set of additive primary colors require specific spectral responses in the camera channels (see Section 2.2).

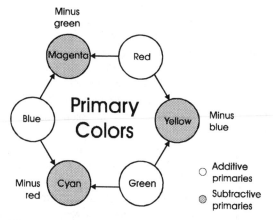

Figure 1.9 The RGB additive primary colors and their complements.

1.8 VIDEO STANDARDS

There are many video system standards in use and no attempt is made here to cover them all or to provide much detail. Such information can be found in [4]. However, the major systems are summarized in this section, with comments relating to how they affect camera design.

1.8.1 Analog Standards

Until the last several years, all video systems were analog. Some digital hardware is used in analog systems, but the signal interconnections between units are defined in analog terms.

Analog systems are designed for the camera and the display to run in exact synchronism, with signal flowing continuously between them in real time. This is a necessity because analog hardware cannot store any significant amount of signal in real time. Some devices for analog systems, such as frame synchronizers, do store a frame or more of video, but they do it by converting the signal to a digital format for the purpose. This makes such features expensive, so they are not suitable for use outside of a studio environment.

1.8.1.1 Component Video

A video camera has signal channels for each of the primary colors. The simplest form of system brings these out of the camera on three cables and connects them to a display that has RGB inputs. That is called a *component* system. However, for a number of reasons, that is almost never used, and few cameras even have RGB outputs. There are other component formats, particularly those based on luminance and color-difference components, such as YC_RC_B and YUV (see Section 1.8.2.1).

The problem with a system based on components is that the three-cable interface excessively complicates any system of multiple cameras, switching, recording, and display. Further, and nearly fatal, is that transmitting or broadcasting component signals requires three channels, which is wasteful of bandwidth and expensive.

Since most systems eventually involve transmission or broadcasting, they must combine (encode) the component signals into a single channel. Such an encoded signal is called *composite* and most systems that deliver a composite output have found that it is actually best to create the composite format at the camera and handle the video that way throughout the system. Thus, composite encoding usually occurs within the camera.

Most analog composite systems are based on combining the signals according to their frequency spectral content (see the next section). However, a few systems were developed using a time-division method, which avoids some of the limitations of frequency combining. These systems are called *multiplexed analog component* (MAC) systems and have seen limited use.

1.8.1.2 NTSC Composite Video

The *National Television Systems Committee* (NTSC) existed in the 1950s in the United States to develop the world's first composite color TV system. This was formally standardized by the Federal Communications Commission (FCC) in 1953 and is known as the NTSC system. Three underlying concepts of NTSC are involved in nearly all color TV systems, analog or digital; they are covered here:

- *Reduced color bandwidth*—because visual acuity is poorer for colored objects than it is for gray scale objects, it is possible to reduce the bandwidth for color-only signals. By structuring the three component channels to consist of one luminance and two color-difference signals, the bandwidth of the color-difference signals can be reduced between two and four times compared to the luminance channel. In the NTSC system, the luminance channel has a bandwidth of 4.2 MHz and the color-difference channels have bandwidths of 1.5 and 0.5 MHz. The latter bandwidth becomes acceptable by placing this color-difference channel in the region of colors (blues) for which the eye's acuity is the poorest. This is an extreme trade-off that some of the other analog composite systems found ways to avoid.

- *Reduced sensitivity to interference at high video frequencies*—Studies of the effect of interference to TV signals show that higher-frequency interference is less visible in pictures than lower frequencies. Interference susceptibility is worst for interfering frequencies in the vicinity of the line-scanning frequency and falls off for both higher and lower frequencies. This makes possible the insertion of a modulated subcarrier at the high end of the luminance channel with little effect on the luminance signal.

Table 1.3
System Parameters of NTSC, PAL, and SECAM Composite Color Systems

Item	NTSC	PAL	SECAM
Total lines	525	625	625
Interlace	2:1	2:1	2:1
Field rate (Hz)	59.94	50.0	50.0
f_H (Hz)	15,734.26	15,625	15,625
Luminance bandwidth (MHz)	4.2	5.0 or 5.5	6.0
f_{SC} (Hz)	3,579,545	4,433,619	4,250,000
			4,406,250
Chrominance bandwidth (MHz)	$I = 1.3$	$U = 1.3$	$D_R = 1.3$
	$Q = 0.5$	$V = 1.3$	$D_B = 1.3$

- *Frequency interleaving*—The visibility of the modulated subcarrier can be further reduced by choosing its carrier to be an odd multiple of one-half the line-scanning frequency. Since a TV signal in general has a spectrum that is concentrated heavily at the line frequency and its harmonics, this choice of subcarrier frequency results in the luminance harmonics and the subcarrier sidebands falling between each other—this is called interleaving.

The interleaved subcarrier is used in NTSC to carry the chrominance components. Suppressed-carrier quadrature amplitude modulation of the subcarrier carries the two color-difference components. Since color-difference components go to zero when there is no color information, and suppressed-carrier modulation has zero output with zero input, there is no interference whatsoever with monochrome transmission. When there are colors present, the frequency interleaving minimizes interference with the luminance channel. By modulating two phases of the carrier in quadrature, the two color differences are sent on a single carrier.

Without going into further detail, the parameters of NTSC composite video are summarized in Table 1.3.

1.8.1.3 PAL Composite Video

Color TV in Europe was not standardized until a few years after the United States and it had the benefit of the early U.S. experience. Some improvements were developed and incorporated into the PAL standard that is used in all of western Europe except France. Scanning standards in Europe are 625 lines, 50 fields, interlaced and video bandwidths are 5 to 6 MHz. The three concepts mentioned above are incorporated in PAL and, in addition, the modulation of the color subcarrier was changed to alternate the phase of one component on alternate lines. This gives the standard its name: *Phase-Alternating Line* (PAL); it allows the chrominance bandwidth to be the same for both channels, and it reduces the effect of amplitude and phase distortions in transmission or recording that may cause color errors in NTSC systems. PAL standards are summarized in Table 1.3.

1.8.1.4 SECAM Composite Video

France did not adopt the PAL standard with the rest of Western Europe but developed its own system called *Sequential Couleur Avec Mémoire* (SECAM), which addresses the same improvements over NTSC that PAL did, but in a different way. To eliminate the phase and amplitude sensitivity of the color subcarrier, frequency modulation is used for the color-difference signals. The two color-difference channels are transmitted on alternate scanning lines and a one-line memory is used in the receiver to recover full color on all lines. SECAM also is the standard for the countries of the former Soviet Union. A summary of the SECAM standards is in Table 1.3.

1.8.2 Digital Standards

New video systems are being designed around digital standards and all classes of equipment, including cameras, are being made for digital systems. There are two levels of standardization that are important here, scanning and signal processing standards, and signal interface standards. Scanning and signal processing is discussed here, while signal interfaces are covered in Chapter 6.

Because existing video systems are in a transition from analog to digital standards, there are many "digital" composite standards based on digitizing of analog-standard signals. This is important, but the full impact of digital systems is not felt until the new digital component standards are used. Both classes are covered below.

1.8.2.1 Digital Components

The highest-performance digital systems are achieved by digitizing the component signals at the camera. This produces separate bit streams for each of the three color components. As in all digital systems, multiple bit streams are easily merged into a single bit stream by interleaving or packetizing techniques (see Section 1.8.2.3). Thus, the creation of a single digital signal (the equivalent of the analog composite signals) is accomplished

Table 1.4

Sampling Parameters of Component Digital Systems Based on ITU-R Rec. BT.601

Standard	f_S (MHz)	Bits/Sample	Data Rate (Mb/s)
4:4:4	13.5	8	324
4:2:2	13.5/6.75	8	216
4:4:4	13.5	10	405
4:2:2	13.5/6.75	10	270
4:1:1*	13.5/3.375	8	162
4:2:0*	13.5/6.75**	8	162

* Not included in Rec. BT.601.

** Color-difference components are also subsampled 2:1 vertically.

by merging the component bit streams and no trade-off is required as long as the resulting composite data rate can be supported.

The International Telecommunications Union (ITU) has created a world standard for digitizing component video using the existing scanning standards (525 and 625 lines). This is ITU-R Recommendation BT.601-4. It offers options for bits per sample and color formats, but all use the same sampling rate for 525- or 625-line systems. These options are summarized in Table 1.4.

The various standards in Rec. BT.601 are identified by three numbers that refer to the sampling rate for each of the components. A rate number of 4 means that sampling is at the basic rate of 13.5 MHz, while a rate number of 2 means that sampling is at one-half the basic rate. The lower rate is used for reducing the bandwidth of the color-difference components by subsampling. Thus, 4:4:4 means that all components are sampled at 13.5 MHz. In this mode, the components may be either RGB or YC_RC_B. The 4:2:2 mode is used only with the YC_RC_B format.

The table shows two other formats (4:1:1 and 4:2:0) that are sometimes used but are not part of Rec. BT.601. They have greater reduction of color bandwidth and are used in video systems that employ compression (see Sections 1.8.2.3 and 13.4.1).

Two other component systems that are sometimes encountered are YIQ and YUV. YIQ is the component basis of analog NTSC, where I and Q are the color-difference components for that system. YUV is the basis of analog PAL; the PAL color-difference components, U and V, are different from those in NTSC.

Data rates for component digital video are high. However, transmission, recording, and signal processing hardware are available to operate at these rates. Most professional production and broadcast facilities being built today use component digital video.

1.8.2.2 Digitized Composite Video

Digitizing of composite analog formats (NTSC or PAL) seems at first to be a waste of time because the compromises in quality that are inherent in the composite encoding remain in the digitized format. However, this is often done in order to enjoy the benefits of digital memory and recording in otherwise analog systems. It is also a cost saving compared to a full component system because the data rates are significantly lower.

The presence of the color subcarrier in an analog composite signal causes composite sampling rates to be chosen at multiples of the subcarrier frequency, either 3× or 4×f_{SC}. Locking sampling to subcarrier simplifies some of the processing, especially when the digitized composite signal must be decoded to components, which is required for video special effects processing. At the same time, the need to decode for some processes is a disadvantage of digital composite systems.

The preferred composite-video sampling rate is 4×f_{SC} which, at 8 bps, delivers a data rate of 114.6 Mb/s for NTSC and 141.8 Mb/s for PAL. Similarly, 3×f_{SC} data rates are 86 Mb/s and 106.3 Mb/s; this is used in some lower-cost systems.

1.8.2.3 Digital Transmission

When transmitting a baseband analog signal, the bandwidth must be equal to or more than the highest frequency component in the video and the transmission SNR must be equal to or more than the desired SNR for the signal. No trade-off is possible. If the analog video signal is modulated onto a carrier for transmission, the bandwidth in general will increase. High quality baseband video transmission for standard TV formats requires bandwidths of about 6 MHz and a SNR in the range from 40 to 50 dB.

On the other hand, transmission of digital signals allows SNR and bandwidth to be freely traded for one another. If a raw bit stream is transmitted, the bandwidth is roughly one-half the bit rate, and the SNR required is in the range of 8 dB. Error protection coding in the bit stream can render nearly perfect transmission through such a system. Encoding methods for bit streams allow the required transmission bandwidth to be reduced at the expense of requiring higher transmission SNR. These subjects are covered in [5].

An important feature of bit streams is their capability for multiplexing. For example, if three equal-rate bit streams are created, as for RGB component video, they may be combined into one stream simply by taking one piece of data at a time from each stream in sequence. The piece of data might be a single sample value or it might be a much larger block of data. Either way, if the receiver knows how the bit streams have been merged (interleaved), the streams are easily separated by reversing the process.

A more sophisticated and flexible multiplexing method is called *packetizing*. In this case, blocks of data are combined with a standardized header block that identifies the data type and other factors that will assist separation. The data block with its header is a *packet* and packets may be interleaved in any sequence. It is not necessary for the individual stream data rates to be equal or for all packets to even have data at all. For example, diverse data streams at widely different rates, such as audio and video streams, may be multiplexed by packetizing.

1.8.2.4 Compressed Video

The ultimate trade-off of data rate is video compression, which takes advantage of the inherent redundancy in scanned video signals to reduce data rate without significant impairment of picture quality. Some compression is possible without *any* loss of quality, but the amount is limited and most video compression techniques go further and make some compromises in areas that will not be too visible to viewers. Data rate reduction of 10:1 or more is achievable this way. The details of video compression are not covered here; [4] provides a good overview of video compression.

Two compression standards are becoming widely used in video cameras and systems. These standards were developed by the *Joint Photographic Experts Group* (JPEG) and the *Moving Picture Experts Group* (MPEG) of the *International Standardizing Organization* (ISO). These standards use similar approaches for still-picture (JPEG) and motion video (MPEG) compression and provide a wide range of options in terms of picture quality and features.

The basis for both JPEG and MPEG is what is known as *transform coding*. The data is divided into blocks of 8 × 8 pixels which are transformed mathematically by a process known as the *discrete cosine transform* (DCT). This transform converts the 64 pixel values of each block into 64 values that represent the frequency spectral content of that block. This, by itself, is not compression but, in the frequency domain, most of the signal energy is in the lower-frequency components, and higher-frequency components can be quantized more coarsely or even eliminated without much affect on the resulting picture. That is what makes the compression. DCT processing, although complex, is available in integrated circuits, and it is practical to build it into even hand-held cameras.

JPEG still-image compression uses DCT as its prime technique. Motion compression can be done by applying JPEG to each frame of a motion stream, a technique called *motion-JPEG*, but more compression is possible for motion video by using *motion compensation*, which is an additional feature that is part of MPEG compression. This process exploits the similarity between successive frames in a motion sequence. Essentially, frame information is transmitted only when something changes. For a still picture, only the first frame is transmitted, subsequent frames are produced at the receiver by remembering the first frame. If part of the frame has motion, motion compensation will find that and, if the moving object can be reproduced by simply moving something that is already in the previous frame in the receiver's memory, it will perform the move at the receiver. Only the coordinates for the move are transmitted. When new information appears in a frame, that will be transmitted using DCT compression. More discussion of compression is in Sections 4.10 and 13.4.

REFERENCES

1. Luther, A. C., *Principles of Digital Audio and Video*, Artech House, Norwood, MA, 1997, Chapter 5.

2. Watkinson, J., *The Art of Digital Video, 2E*, Focal Press, London, 1994, pp. 139–147.

3. Benson, K. B., and Whitaker, J., *Television Engineering Handbook*, McGraw-Hill, New York, 1992, pp 2.14–26.

4. Luther, A. C., *Principles of Digital Audio and Video*, Artech House, Norwood, MA, 1997, Chapter 9.

5. Ibid., Chapter 7.

2

Light and Optics

The input medium of a video camera is light modulated by being reflected from the scene in front of the camera. An optical system in the camera directs the light to the image sensors in the camera. This chapter discusses the properties of light and the considerations of optical systems for cameras.

2.1 ILLUMINATION

The light source for a natural scene is its *illumination*. As with any light, illumination has the properties of intensity and color, which significantly affect the performance of the camera. Typical illumination sources are the sun (outdoors), incandescent or fluorescent lights (indoors), and others. Figure 2.1 shows a scene with its illumination.

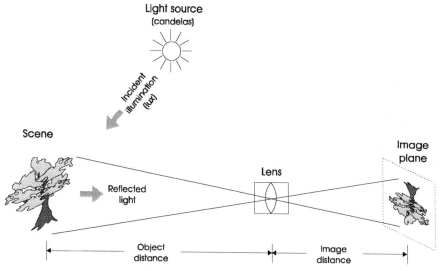

Figure 2.1 Illumination.

Table 2.1
Illuminant Intensity of Various Scenes

Scene	Typical Illumination (lux)
Bright sunlight	100,000
TV studio	2,000
Street scene at night	20
Scene lighted by 1 candle	1
Full moon illumination	0.1

2.1.1 Illuminant Intensity

As with any source of radiant energy, an illumination source can be characterized in terms of energy density (W/m^2). However, there are special measures for light intensity that consider the nature of light and its use. In illumination situations, light is measured in terms of how it looks to a human viewer, so it is convenient to characterize light intensity by taking into account the luminosity curve (see Section 1.4.2). An international unit for this is the *lumen*, which is the time rate of flow of *luminous intensity*. The lumen is defined in terms of fundamental physical properties.*

Since the light energy from a source spreads out as one moves away from the source, the actual amount of energy falling on an area of a scene changes as the inverse square of the distance between source and scene. To account for this, incident light on a scene is measured as lumens per unit area: one lumen per m^2 is defined as 1 *lux*. In English units, one lumen per ft^2 is a *foot-candle*. Thus, an illumination of 1 lux = 0.093 foot-candles. The light sensitivity of video cameras is usually expressed in lux but, since the camera responds to reflected light, the reflectance of the scene must also be specified.

The illuminant intensity of a variety of scenes is listed in Table 2.1. Video cameras are expected to operate under any of these conditions, which poses a major challenge to camera designers.

2.1.2 Illuminant Color

Generally, illumination should be white, which means it includes a broad spectrum of colors. Because of the ability of human vision to adapt to the illuminant, there are many colors that appear "white" once our eyes have had a few seconds to adapt. Colors that appear white after adaptation by most observers is shown on the CIE chromaticity diagram (Figure 1.7) as an area at the lower center of the horseshoe. Exact specification of an illuminant is done by chromaticity coordinates x and y, but such numbers are not easily interpreted in terms of "color."

---

*   A light source emitting energy uniformly in all directions with an intensity of 1 *candela* emits 1 lumen per steradian (unit solid angle.) One candela is defined as the luminous intensity of 1/600,000 m^2 of projected area of a blackbody radiator operating at the temperature of solidification of platinum and a pressure of 101,325 n/m^2.

Table 2.2
CIE Standard Illuminants

CIE Illuminant	x	y	Use
A	0.4476	0.4074	Tungsten incandescent
B	0.3484	0.3516	Noon sunlight
C	0.3101	0.3162	Average daylight
D_{65}	0.3127	0.3290	Daylight

Another illuminant specifier is *color temperature*, which is less precise than chromaticity coordinates, but easier to understand. Color temperature is the temperature of a blackbody radiator that produces a matching visual sensation to the illuminant. For example, incandescent illumination has a color temperature in the range of 3,000K. Daylight usually is defined as a color temperature of 6,500K, and so on. The locus of color temperatures shows on the CIE diagram as a line beginning at the red end of the spectrum for low color temperatures, and curving out toward the center of the diagram for high temperatures (see Figure 1.8). Except for colors that are exactly on this line, color temperature is only an approximate specification for an illuminant.

Video cameras have no natural ability to adapt to the illuminant. They must be told what color in the image to make "white" and then a *white balancing* procedure must be performed. This can be automated, as explained in Chapter 5.

Certain illuminants are standardized by the CIE. Names, chromaticity coordinates, and uses for these are tabulated in Table 2.2. For most purposes in video systems, illuminants A and D_{65} are sufficient.

2.1.3 Reflection and Refraction

Light impinging on transparent materials may undergo reflection or refraction.

The reflection process is visualized in Figure 2.2. *Reflection* may occur at any interface between materials of different *refractive index*, which is the ratio of the free-space speed of light c to the velocity of light propagation in the medium. An incident ray is reflected at the interface, coming off at an exit angle to the normal that is a mirror of the angle of incidence θ_1. A portion of the incident ray may also be *refracted* into medium 2 at an angle θ_2 calculated by:

$$n_1 \sin \theta_1 = n_2 \sin \theta_2 \tag{2.1}$$

where n_1 and n_2 are the refractive indices of medium 1 and medium 2, respectively.

A special situation, known as *total reflection* can occur at the interface with a less-dense medium, when the angle of incidence is greater than a critical value θ_C:

$$\theta_C = \sin^{-1}(n_1/n_2) \tag{2.2}$$

For example, total reflection is the property that allows light waves to stay inside an optical fiber.

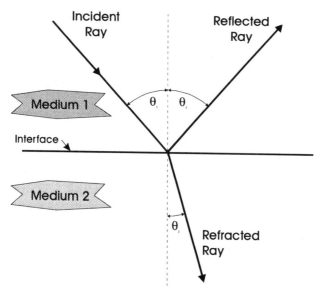

Figure 2.2 *Reflection and refraction.*

2.1.4 Scene Reflectance

A camera does not see the incident illumination on the scene—it sees the *reflection* of the illumination by each area of the scene. The reflection process modifies the intensity and spectral distribution of the illumination because of the surface properties (reflectance, texture, and so forth) of each object in the scene. This gives color to the scene even though the illumination is "white."

Reflection from diffuse surfaces, as exhibited by most objects in a natural scene, behaves differently from the model described in the previous section. Light is scattered equally in all directions, as described by *Lambert's cosine law*

$$I_D = I_P k_D \cos \theta \tag{2.3}$$

where I_D is the reflected illumination from a point source of intensity I_p located at a position, which is at an angle θ from a normal to the reflecting surface, and k_D is the diffuse-reflection coefficient of the surface, having a value between 0 and 1.

This is independent of the viewer's position, so an object appears the same brightness from any viewing location.

A shiny surface that is capable of normal reflection will reflect a point-source illumination directly into the camera only when the camera and light source bear the relationship shown in Figure 2.2. For example, a shiny red ball will have a *specular* highlight on its surface only at the position where the ball's curved surface meets the normal reflection condition. The highlight has the same color as the illuminant (white) whereas, at all other positions on the ball, the reflection is diffuse and appears red.

Except for specular objects, scenes never have a maximum reflectance of 100 %. The usual method of specifying and measuring camera sensitivity calls for a test chart that has a maximum reflectance of 89.9 %, which corresponds to white video level (see Chapter 12).

2.1.5 Absorption and Transmission

Materials that transmit most of the light they receive are said to be transparent. Materials that do not transmit very much light are opaque. There is a continuum between these limits and the transmission properties are also often a function of wavelength. Incident light energy that is not reflected or transmitted is absorbed within the material. These properties of absorption are useful in optical filters (see Section 2.6).

2.2 VIDEO COLORIMETRY

The science of color matching is *colorimetry*.* In video cameras, colorimetry involves the specification of spectral responses for each of the RGB camera channels (the *camera taking curves*) and the signal processing that may modify electronically the effect of the spectral responses. Electronic modification of camera taking curves, usually by linear matrixing among the RGB signals, is called *masking*.

2.2.1 Camera Taking Curves

In an additive color reproduction system, the camera taking curves are calculated from knowledge of the exact primary colors to be used by the display devices. The display primaries must be standardized, which is done by specification of chromaticity coordinates for the primary colors. Actual calculation of the taking curves will not be detailed here, but it is a straightforward process of matrix arithmetic [1]. The calculation also must consider the illuminant; most calculations are done today with illuminant D_{65}. Figure 2.3 shows the taking curves for an NTSC television camera.

The curves have been normalized so their areas are equal. Notice that these curves have negative lobes. Practical imagers have only positive output signals and cannot produce negative lobes. The necessary negative responses are produced in the camera's signal processing by masking. For example, the negative lobe of the green channel around 450 nm is obtained by subtracting a small amount of the blue signal from the green channel.

The taking curves result from combination of the spectral responses of the camera

---

* Colorimetry originally meant the measurement of color with a colorimeter. Today, it has the broader meaning of the science of color.

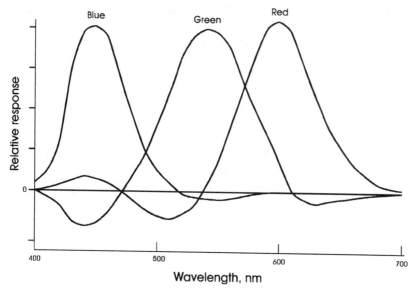

Figure 2.3 *NTSC color camera taking curves.*

light-splitting optics, the image sensor, and any optical color filtering included in the camera. Masking is used to touch up the resulting response and add the negative lobes.

Measurement of taking curves for a complete camera is done using an adjustable monochromatic light source that delivers a calibrated intensity of any spectral color. This is a difficult measurement that is usually done by a camera manufacturer at the time of design or manufacture. It is seldom done by camera users. Taking curves should remain stable over time if the camera optics remain clean. An important aspect of optics design is to prevent accumulation of optical dirt or contamination that could affect camera performance.

2.3 LENSES

The camera lens is the major optical component and is usually the most expensive optical item. It determines the camera's focusing capability and contributes to resolution, sensitivity, and color performance. Lenses make use of refraction at curved surfaces, usually glass. An excellent overview of optical systems for video is published by Canon [2].

2.3.1 Lens Theory

The science that supports lens design is *geometric optics*. No attempt is made here to cover that subject fully, but certain features are discussed to help the reader understand how lenses behave and how the various design parameters affect lens performance. Further information is available in reference [3].

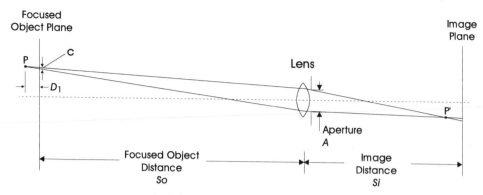

Figure 2.4 *Geometry of thin lens operation.*

2.3.1.1 Thin Lenses

Figure 2.4 shows the operation of a simple lens, where the thickness of the lens may be neglected compared to the focusing distances. This is referred to as a *thin lens*. Scene objects located in the *object plane* at the left are focused by the lens onto an *image plane* at the right. In a video camera, the image sensors are placed so their sensitive surfaces are in the image plane. The relationship between the object distance S_O and the image distance S_I depends on the lens *focal length f* according to

$$1/S_O + 1/S_I = 1/f \tag{2.4}$$

The focal length is the distance between the center of the lens and the image plane when focusing objects at an infinite distance. It is determined by the curvature of the lens surfaces and the refractive index of the lens material. Most thin lenses have spherical surfaces because that shape is easily manufactured.

Figure 2.4 also shows how an object P located behind the object plane (farther from the lens) is focused in front of the image plane at P′. The cone of light rays from P is enlarged at the image plane itself, resulting in a slightly out-of-focus image. This is called the *circle of confusion*, which is shown by the area C in the object plane.

The circle of confusion leads to the concept of *depth of field*, which is the range of distances in the object plane that appear to be in focus at the image plane. The words "appear to be in focus" are key here—objects are "in focus" for a video camera as long as the image sensor cannot detect the defocusing. That is determined by the resolution of the image sensor; the light cone from a point appears to be a point if it spreads no more than one pixel size at the image plane.

The light-gathering ability of a lens depends on its diameter; a larger diameter collects more light. However, for a given focal length, a larger-diameter camera lens causes less depth of focus. It is often desirable to adjust the effective lens diameter by placing an adjustable aperture (an *iris*) in the light path near the lens.

Table 2.3
F-Numbers versus Relative Sensitivity

F-number	2	2.8	4	5.6	8	11	16	22
Relative sensitivity	128	64	32	16	8	4	2	1

The effect of an aperture may be normalized by relating its diameter A to the lens focal length, giving rise to the *f-number*

$$\text{f-number} = f/A \tag{2.5}$$

The minimum f-number of a lens is an important specification. The smaller the f-number, the "faster" the lens, meaning it collects more light and it will make a camera more sensitive. Since light-gathering ability depends on the *area* of the aperture, sensitivity changes as the square of the f-number. This is shown in Table 2.3. Of course, f-numbers could have any values, but the values shown are the most commonly used for photographic and video lenses. Specific f-number settings are called *f-stops*, closing the aperture of a lens is called *stopping down*.

2.3.1.2 Thick Lenses

Most practical lenses require more than one optical element and the simplification of thin lenses may not be appropriate. This is accommodated by the concepts of *thick lenses*. Figure 2.5 shows the case of a thick lens. Two planes in the lens are required for the definition of object and image distances; these are called the *principal planes* of the lens. By measuring S_O and S_I from the principal planes, (2.4) still applies to thick lenses.

Notice that the principal planes may be well inside the lens for some designs. This means that the distance from the lens surface to the image plane (the *back focal distance*)

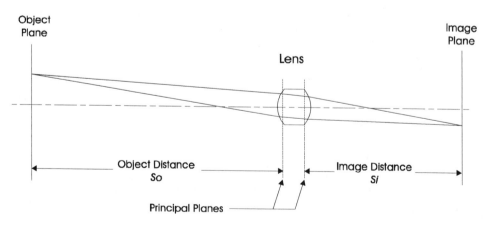

Figure 2.5 *A thick lens.*

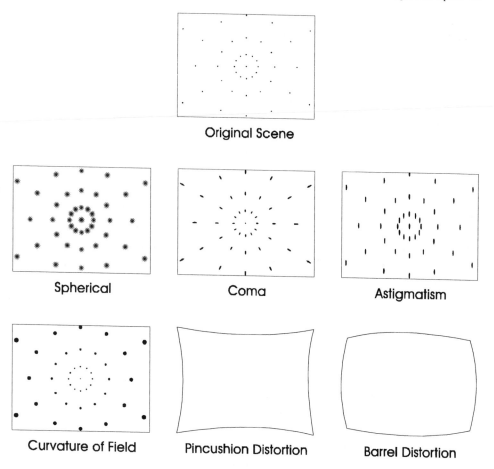

Figure 2.6 *Aberrations. The lens is adjusted for best focus at the center of the image.*

may be less than the focal length at infinity focus. When a camera uses a light-splitter for color separation, the back focal distance must be greater than the optical path length through the light-splitter, or the camera will be unable to focus to infinity.

2.3.2 Lens Aberrations

Lenses do not necessarily provide perfect reproduction of the object at the image plane. Certain distortions occur in lens construction—these are called *aberrations*. There are five principal types of aberration for monochromatic (single wavelength) light; spherical aberration, coma, astigmatism, curvature of field, and distortion, shown exaggerated in Figure 2.6. In addition, there are chromatic aberrations when the light covers a range of the visible spectrum. Aberrations can be corrected by the appropriate combinations of multiple lens elements. The aberrations are discussed in this section.

2.3.2.1 Spherical Aberration

Spherical aberration is caused by light rays that pass through the edge of the lens focusing at a different distance from the lens than rays passing near the axis of the lens. This aberration is reduced by stopping down the lens with an aperture. It is also correctable by combining lens elements having equal but opposite spherical aberrations.

2.3.2.2 Coma

In spherical aberration, rays are focused at different distances along the axis depending on where they pass through the lens. Coma is a sort of perpendicular aberration in that rays focus at different distances off the axis depending on where they pass through the lens. The result is that a point object focuses to a comet-like shape in the object plane. Coma can be eliminated in a single lens at a given object and image distance by properly choosing the radii of the lens surfaces.

2.3.2.3 Astigmatism

Astigmatism combines elements of both spherical aberration and coma in that a point is focused to an ellipse in the object plane. As the focus is adjusted, the image changes between two perpendicular ellipses, but never reaching a point of sharp focus.

2.3.2.4 Curvature of Field

When off-axis objects focus to different image planes along the axis, with the plane positions depending on the distance off-axis, it is curvature of field.

2.3.2.5 Distortion

If a lens has different magnification corresponding to the distance off-axis, it is a case of distortion. If the magnification increases for off-axis objects, it is *pincushion* distortion, if magnification decreases, it is *barrel* distortion.

2.3.2.6 Chromatic Aberration

The previous five aberrations occur with monochromatic light. With colored or broad-spectrum light, two additional *chromatic aberrations* occur. *Axial* or *longitudinal* chromatism occurs when the position of the image plane moves along the axis with wavelength. This is effectively a change of focal length with wavelength and thus, a change of magnification. The result is *lateral* chromatism, where the images of different colors are not the same size.

2.3.3 Lens Design Issues

Lens designers must deal with the aberrations mentioned above as well as many other factors that become important in video camera lens application. These factors are discussed in this section.

2.3.3.1 Image Diagonal

Lenses are designed for a specific size of image. In a video camera, that is the size of the imager, specified as a diagonal measurement, usually in inches. Typical sizes are 2/3, 1/2, and 1/3 in. (Metric values are sometimes used; the corresponding values are 16, 11, and 9 mm, respectively.) The smallest imagers produce the smallest camera, but small imagers also have less resolution because of practical limits to pixel size in fabricating the imagers. Thus, higher-performance cameras use larger imagers.

A larger image size implies longer focal length lenses for the same angle-of-view performance. The relationship is

$$\theta = 2 \tan^{-1} (d/2f) \tag{2.6}$$

where θ is the angle of view;
  d is the image diagonal;
  f is the focal length.

This is expressed graphically in Figure 2.7 for common image sizes. There is an awkwardness of units in this figure because image sizes are usually given in inches, but focal lengths are always specified in millimeters. However, in (2.6), both d and f must be specified in the same units. A "normal" angle of view is 50 degrees; lenses with larger angles are *wide-angle*, and those with smaller angles are *telephoto*.

2.3.3.2 Lens Resolution

The resolution of a lens is theoretically limited by diffraction. Practical lenses are further limited by aberrations (see Section 2.3.2). The diffraction-limited resolution is highest when the lens is wide open (lowest f-number) and reduces as the lens is stopped down. This is not usually a problem for lenses operating in 525- and 625-line video systems, but it becomes an important lens design issue in lenses for HDTV (see Section 13.3.3).

2.3.3.3 Vignetting

A given lens design has a limit to the maximum angle of view because, at wide angles, the lens mounting structure may interfere with some of the light rays. That causes a loss of image brightness in the corners, which is called *vignetting*. This effect and certain aberra-

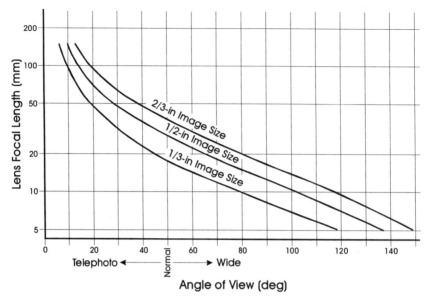

Figure 2.7 *Lens focal length versus angle of view.*

tions prevent using a lens for larger image sizes than its design value.

2.3.3.4 Zoom

The technology of variable focal length (zoom) lenses has advanced to the point that virtually all video cameras from home units to broadcast and production cameras use zoom lenses.

Zooming is achieved by using multiple lens elements that are moved relative to one another to vary the focal length of the combined lens. A typical zoom lens design is diagrammed in Figure 2.8. Zooming is accomplished by moving a lens group called the *variator* along the optical axis. When the variator is close to the front of the lens, it gives a wide angle; when it is close to the back of the lens, it gives a telephoto. Because the zooming action affects focusing, a second moving lens group, called the *compensator*, tracks the variator to keep the lens focused at a fixed object distance. The motions of the variator and the compensator must be precisely coordinated, which is done by a cam mechanism designed into the barrel of the lens. Focusing of the lens to different object distances is done by moving the front lens elements (the focusing group). It is extremely challenging to achieve a zoom lens design that will allow control of aberrations over the entire zoom range. However, using computer synthesis and other advanced tools, zoom ranges up to 70:1 have been developed.

Since the lens size and cost grows rapidly as the zoom range is increased, the longest ranges are available only in lenses for the largest, most expensive cameras. Low- and medium-priced cameras have zoom ranges from 10:1 to 15:1, which is achievable with-

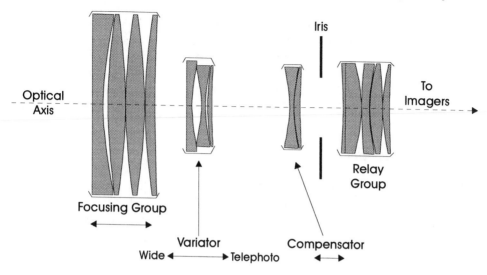

Figure 2.8 *Diagram of a zoom lens.*

out excessive size and cost.

Zoom lenses are specified by their range and minimum focal length; for example, a 10×6.5 lens has a 10:1 range and a minimum focal length of 6.5 mm—it zooms from 6.5 to 65 mm.

Control of the zooming is either manual, by means of a ring or lever on the lens itself, or remote-controlled with an electric motor and switches. In the latter case, many cameras have variable-speed zoom control to give the maximum flexibility in use. This is a more expensive feature that is not generally available in the lowest-priced cameras (see Section 5.6).

2.3.3.5 Aspherical Lenses

Historically, lenses have been made with spherical surfaces because they were easily manufactured. However, spherical lenses have inherent aberrations that require multiple lens elements for correction. This problem can be eliminated by use of a different surface shape. Recent advances in lens manufacturing technology have made possible economical production of *aspherical* lenses that have much reduced inherent aberration. This allows higher-performing lenses to be made with fewer elements, reducing size, weight, and cost.

2.3.3.6 Internal Focusing

Most lenses are focused manually by rotating a ring on the outside of the lens. This action moves the lens elements to adjust focusing, but it is awkward in that the entire front of the lens rotates, and any filter attached to the front of the lens also rotates. It also exposes moving parts to dirt and dust. Another recent advance is the *internal focus* lens, which embeds all the moving parts within the lens or camera. This requires an internal motor to accomplish the movement, but that is required anyway if the camera has automatic focusing (see Section 5.2). For manual focusing, the camera must have external switches that control the internal motor. This feature adds to the cost, so it is usually not on the lowest-priced cameras.

2.3.3.7 Macro Capability

Another important lens feature is the ability to focus at short distances to achieve extreme close-up pictures. This is known as *macro* capability. Some lenses can focus as close as ¼ in. However, macro operation on most lenses requires the zoom to be set at wide angle, and the camera lens may cast a shadow on the scene. A rarer capability is to have macro focusing at any zoom setting.

2.3.3.8 Halation

As light passes through a lens, spurious reflections or scattering may direct light away from the optical path and thus, reduce the light transmission of the lens. This is proportional to the number of elements in the lens, although there are some design features to minimize it.

It is important that spurious reflections in the lens or camera optical system be absorbed and not be allowed to return to the optical path. The latter effect, if focused, causes a ghost or reflection in the picture. If internal reflections returning to the optical path are not focused, they cause a general loss of contrast in the picture. This is called *halation* and may be an overall loss of contrast or a local effect that surrounds highlights in the image. In either case, it is undesirable.

Lens elements are usually coated with a special surface to avoid reflections. In addition, newer lenses may use *low-dispersion glass* that reduces scattering and transmission losses.

2.3.3.9 Lens Mounts

Before the days of zoom lenses, video cameras had interchangeable lenses to obtain different focal lengths. A standardized lens mount was an important feature to support interchangeable lenses. Since most cameras now have zoom lenses, the need for interchanging lenses is reduced. However, especially in professional cameras, there is still a need, and several standard mountings are used. This allows trading lenses (e.g., between an extra-wide range zoom lens, which may be too large and heavy to use all the time, and a lesser-range lens for general use.)

Bayonet mounts require only a quarter-turn twist to attach or remove a lens. This is quick and easy, and there are several varieties of bayonet mounts for different image sizes and manufacturers. Another mounting involves threads to screw the lens onto the camera. The C mount is a variety of this that was used for many years. Newer cameras favor bayonet mounts.

2.4 LIGHT SPLITTERS

Color cameras require separation of the incoming light into the three primary colors to produce the proper taking curves for color reproduction (see Section 2.2.1). In single-imager cameras, light splitting occurs right in the imager, but three-imager cameras require separate light-splitting optics to direct the colors spatially to three individual imagers. Single-imager cameras are discussed in Section 3.5; three-imager light splitting is discussed in this section.

2.4.1 Dichroic Mirrors

The objective of color light splitting is to separate the color components according to the camera taking curves while losing as little light energy as possible. The *dichroic mirror*, shown in Figure 2.9, is the key to efficient light splitting. It allows certain wavelengths to pass through and other wavelengths to be reflected, without any absorption of light.

A dichroic mirror consists of one or more thin coatings on an optical substrate, such as glass. The key is that the coatings have different refractive indices from the glass, and precise thicknesses. Interference between the light waves reflected or refracted at each surface in the filter causes a wavelength-dependent behavior as shown in Figure 2.9(b). Blue light is reflected by the first mirror but longer-wavelength light passes through to the second mirror. That mirror is designed to reflect red light, but green light passes through. Thus, the light is split into its three components.

Dichroic mirrors perform rough light splitting by wavelength, but the final spectral response must be trimmed by adding color filters in each channel to achieve the exact taking curves. This is evident in Figure 2.9(b), where the green response is nearly produced by the dichroics alone, but the red and blue responses need trimming to reduce the response at long and short wavelengths. An objective of light splitter design is to minimize the amount of trim filtering because it can cause poorer light transmission.

2.4.2 Prism Optics

Although the dichroic mirror light splitter shown in Figure 2.9 does the basic job, it has a number of undesirable properties from the viewpoint of manufacturing and operation of cameras. It can be seen that the optical path through the splitter is quite long, which calls for a long back-focal distance lens or, as was done in early cameras that used mirrors, a *relay lens* is needed to deliberately lengthen the optical path. This is expensive.

A more important problem with mirror optical systems is that the mirror surfaces are exposed to atmospheric contamination and possible degradation. Further, the separate

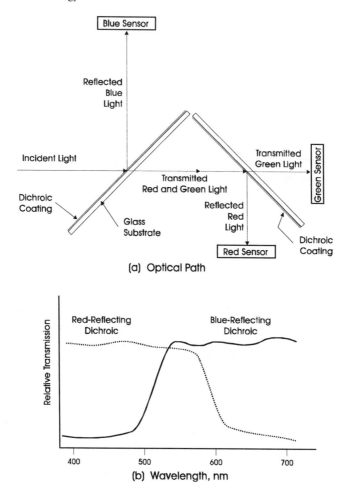

Figure 2.9 *Dichroic mirror light splitter.*

elements of the splitter must be placed precisely and maintained that way, which is difficult in a large assembly of individual pieces. All these problems are solved with *prism
optics*. The dichroic coatings are placed on the surfaces of several solid glass prisms,
which are permanently cemented together along with the imagers as shown in Figure
2.10. This creates a reliable assembly with a minimum optical path length.

An optical path in glass is effectively lengthened compared to the physical path by the
refractive index of the glass. That allows prism optics to use shorter back focal length
lenses.

Three glass prisms A, B, and C are separated by air gaps. A fourth prism D is cemented to prism C. Prisms B, C, and D carry the three imagers on their outer surfaces.
Trim filters as needed are placed between the imagers and the prism surfaces. Prisms B
and C have blue-reflecting and red-reflecting dichroic coatings, respectively, on their

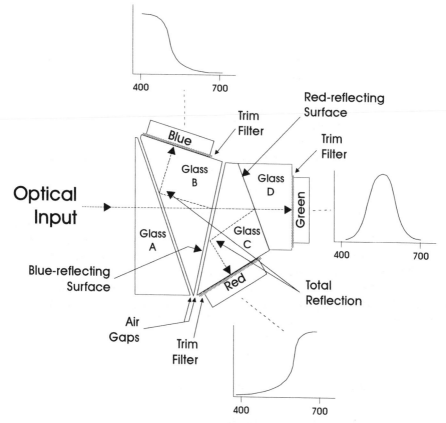

Figure 2.10 *A prism light splitter for an RGB color camera.*

surfaces farthest from the optical input. The figure shows how the blue and red light is reflected to the appropriate imagers; green light passes straight through the prism assembly to the green imager.

The purpose of the air gaps in the assembly is to cause total reflection of the blue and red light inside their prisms. It is important to control the angle of light rays passing through the prisms to not violate the total-reflection condition. A problem may occur with this when using lenses with very low f-numbers. Prism A is added to the structure to shift the angles of light passage to avoid nontotal reflection. It also serves to protect the dichroic coating on prism B, to provide a perpendicular surface for the optical input port, and to fill the rest of the optical path with glass.

The prism structure is designed for equal optical path length to each of the imagers, and in most cameras, the entire assembly is permanently cemented together during manufacture and never requires adjustment in use. Prism optics were a major advance in camera technology when introduced and they continue to be the preferred approach for three-imager cameras.

2.5 OPTICAL FILTERS

An ideal optical system provides maximal transmission at all wavelengths. However, there are situations where reduced transmission is desired or transmission must be changed at different wavelengths. Both of these are applications for *optical filters*, which are discussed in this section.

Many filter effects can also be accomplished by electronic processing in postproduction, but this is expensive. If a simple filter costing only a few dollars and placed over the lens can accomplish a desired effect, it is much faster and cheaper than video processing. In other cases, a filter helps the camera capture the scene without distortion, such as when the scene contrast is too high or the scene is too bright.

2.5.1 Neutral-Density Filters

A filter that has the same transmission factor at all wavelengths is a *neutral-density filter*. Such a filter appears gray to the eye. Most cameras have some means to introduce neutral-density filtering when there is too much light for the camera even when the lens is stopped all the way down. It also may be used in a normal lighting situation to force the lens to a wider aperture, thus deliberately reducing the depth of field. This is often done in cases where a background needs to be kept out of focus.

2.5.2 Color Filters

Color filters are used in cameras to shape the taking-channel spectral responses as described in Section 2.4. They also are used for rough color balancing of a camera for different illuminations, or to provide various special effects with color.

As was just described in Section 2.4.1, color filtering may be done with dichroic coatings. It also is done by using the absorption properties of certain chemicals. These result from the specific particle structure of the material that causes optical interference of certain wavelengths as light passes through the filter. Certain wavelengths are absorbed and others are passed. With some light being absorbed by the filter, of course, the camera sensitivity will be reduced. Since extreme coloring is rarely used, the loss of sensitivity with color filtering is seldom important.

Filters for color balancing are built into most cameras and are selected by means of an external control switch or wheel. They are usually placed on a filter wheel that can rotate different filters into the optical path just ahead of the prism input port. Often, there are two filter wheels to support independent selection of color-balancing and neutral-density filters.

2.5.3 Polarizing Filters

A property of light that has not yet been mentioned is *polarization*. This results from the wave nature of light, which may be represented as oscillating electric and magnetic field vectors in a plane perpendicular to the direction of propagation. In general, the electric

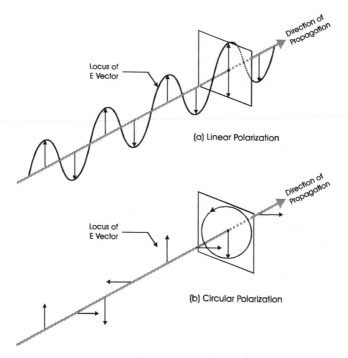

Figure 2.11 (a) Linear polarization, (b) circular polarization.

field vector may move at an arbitrary angle in its plane—its tip traces an ellipse. However, by certain means, the oscillation of the electric vector may be constrained to a straight line—this is *linear* polarization, shown in Figure 2.11(a). If the electric field vector traces a circle in its plane, it is *circular* polarization, Figure 2.11(b). Many optical devices use the polarization properties of light.

Light from a natural scene that has experienced scattering generally is nonpolarized. However, a *polarizing filter* can eliminate a specific plane of polarization. Such a filter is made from layers of transparent crystalline materials that have different velocities of propagation for light in different directions. This is equivalent to a different refractive index for different directions.

A polarizing filter receiving a nonpolarized input delivers a linear polarized output in a specific direction. If the filter is rotated about the optical axis, the polarization plane of its output will rotate correspondingly. Similarly, if two polarizing filters are placed in the light path and rotated to perpendicular directions of polarization, all light will be removed by the filters.

Polarizing filters are useful with cameras for removing glare or bright reflections from shiny surfaces, which tend to reflect light of one predominant polarization. By rotating the filter to reject the predominant polarization, glare is removed without affecting the rest of the scene. Polarizing filters generally are placed in front of the camera lens where they are readily accessible for adjustment.

2.5.4 Specialty Filters

Filters provide many special effects, sometimes easier and less expensively than electronic signal manipulations, which is another popular way of generating special effects. Some of the most important filter-based effects are discussed in this section. As with polarizers, most special-effect filters are placed in front of the lens rather than inside the camera, so they can be easily changed.

2.5.4.1 Graduated Filters

When a filter has varying color across its surface, it is graduated. A filter that is blue at one edge and changes to clear at the opposite edge is used to enhance the blueness of sky in landscape shots without changing the colors of the landscape itself. This works because the light cone in front of the lens is partially focused so that light from the top of the scene passes mostly through the top of the filter. Similarly, light from the bottom of the scene passes mostly through the lower part of the filter. Various other effects are possible with graduated filters, generally dealing with a gradual shifting of color or brightness across the entire scene.

2.5.4.2 Contrast Filters

A filter that scatters light back into the optical path reduces the contrast of the picture. This may be used to deliberately reduce scene contrast for better reproduction by the limited contrast range capability of a video system or it may be used to create a special effect, such as fog. Contrast-reduction filters can operate either by adding light to the dark areas of the scene (called *low-contrast* filtering) or reducing the highlight brightness (called *soft-contrast* filtering.)

2.5.4.3 Diffusion Filters

Sometimes, it is desired to make a scene deliberately soft; this is an application for a *diffusion filter*. These filters have a light-scattering pattern that softens sharp edges in the scene, even when the lens is in sharp focus. The effect is different and generally more pleasing than simply defocusing the lens. Often defocusing the lens on a foreground object will bring some other part of the scene into focus; a diffusion filter will soften the foreground without sharpening any other part of the picture.

2.5.4.4 Dramatic Effect Filters

There are a host of other filters that produce unusual special effects. The star filter has one or more grating patterns that cause highlight points to be stretched into lines or stars. Rotating the filter rotates the star effect. Another special effect is deliberate vignetting to produce an image that appears to be seen through a soft-edge hole. This is produced by a filter that is clear in the center and diffused toward the edges. There are many other possibilities.

REFERENCES

1. Benson, K. B., and Whitaker, J., *Television Engineering Handbook*, McGraw-Hill, New York, 1992, Section 2.4.

2. Canon, Inc., *TV Optics II: The Canon Guide Book of Optics for Television Systems*, 1992.

3. Benson, K. B., and Whitaker, J., *Television Engineering Handbook*, McGraw-Hill, New York, 1992, Section 3.1.

3

Image Sensors

The heart of a video camera is its image sensing, which performs the fundamental task of converting optical images to electrical signals. In doing this, it combines the technologies of photodetection and scanning. Image sensors or *imagers*, specifically *charge-coupled devices* (CCDs), are the subject of this chapter.

3.1 FUNDAMENTALS

An imager can perform the steps of photodetection and scanning in either order, but there is a massive advantage to converting an optical image to an electronic image *before* scanning. That is the only way an imager can make use of all the light energy in the optical image (see Sections 3.1.2 and 3.1.3).

3.1.1 Photodetection

Many materials are capable of converting optical energy (light) into electrical energy. There are several different methods:

- *Photovoltaics*—This is the technology of solar cells used for power generation. An electrical voltage is generated proportional to the illumination on the surface of the cell. Although this is effective for power generation, it is not applicable to imagers.

- *Photoemission*—In this method, the material (usually a metal alloy) emits electrons from its surface when illuminated. This is useful only in a vacuum, but it is the basis for many of the tube-type imagers used in the early days of TV.

- *Photoconductivity*—A semiconductor material increases in conductivity when illuminated, This is because the light energy causes free electrons (and holes) to be generated in the material. If an electric field is impressed on the material, a current (*photocurrent*) flows proportional to the light intensity. Modern tube-type and all CCD imagers use photoconductivity.

3.1.2 Instantaneous Imaging

Before the days of vacuum tubes, there were many attempts to build imagers by mechanically scanning a single photodetector across an image. The most successful of these was the *Nipkow disk*[1] developed in 1884 by Paul Nipkow, a German engineer. This device uses a rotating wheel containing a series of small holes or lenses arrayed in a spiral around the center of the disk—one for each horizontal scanning line. The disk rotates at frame rate in front of a single photodetector. An image of the scene is focused onto the disk, and the disk's lenses direct a point from that image to the surface of the photodetector; the location of the point in the scene scans as the disk rotates.

The result is *instantaneous* imaging because the light energy at each point in the scene is measured instantaneously as the scanning location passes over that point. At all other times, the light coming from the image to that point is thrown away. This produces a signal, but the sensitivity is extremely poor, and workable instantaneous imagers require a very bright scene. In spite of that, instantaneous imagers are used today for certain special purposes, such as telecine cameras (for capturing video from motion picture film, see Section 7.2), and in some character recognition systems.

3.1.3 Storage Imaging

By forming an electronic image *before* scanning, all the light at each point in the image can be integrated over the entire frame period. This is done by storing a charge image resulting from the photocurrent or photoelectrons being generated at each point. A separate scanning process reads out the charge image as an electrical signal. Compared to instantaneous imaging, the improvement in light energy collection is approximately equal to the number of pixels in the image (hundreds of thousands)—a massive advantage indeed. All imagers designed for video capture from live scenes use storage.

In tube imagers, storage occurs directly in the sensitive material, which may be a semiconductor that simultaneously performs photoconductivity and acts as a capacitor to store charge. This requires a very thin coating so that the stored charge does not spread laterally too much in a frame time, because that would blur the image. In a tube imager, the charge image is read out by scanning it with an electron beam.

In solid-state imagers, a rectangular array of separate cells (pixels) is produced by the manufacturing process. Sensing and storage take place in each cell of the imager and specific structural features prevent any spreading of charge from one cell to another. Solid-state imager structures also contain means for reading out the charge from the cells according to a specified scanning pattern.

3.2 CHARGE-COUPLED DEVICES

Although tube-type imagers served for the first 40 years of television, they have now almost completely been replaced by solid-state *charge-coupled device* (CCD) imagers. Through advances in solid-state processing technology, current CCD imagers have been developed to the point where they offer advantages over tubes in almost all respects. The following list covers the features of CCDs.*

- *Size, weight, and power*—CCD imagers are small and power-efficient. Because of this advantage, they were first introduced in hand-held or portable cameras. When the other performance factors were perfected, CCDs became used even in cameras where size, weight, and power are not the primary concerns.

- *Sensitivity*—Greater quantum efficiency in the conversion of light energy to electrical charge, better spectral response, elimination of lag (see Section 3.4.7), and better SNR leads to as much as 4:1 improvement in CCD camera sensitivity.

- *Electrical and mechanical stability*—A CCD is a solid-state device whose picture geometry is determined by the precise patterns etched, diffused, or deposited on the solid-state substrate. The result is an essentially monolithic structure, which gives complete mechanical and electrical stability of the image. This allows the CCD and its optics to be cemented together at the factory, eliminating the need for field adjustment of geometrical or electrical parameters to compensate for drift.

- *Highlight handling*—Except under the most carefully controlled studio conditions, real images always contain *highlights*. These are areas or points in the scene that have a light level significantly higher than the "white" level in the desired areas of the scene. Highlights can be 10 or more times higher than white level in other areas of the scene, especially when a direct reflection (called a *specularity*) of the illumination occurs. Through the combination of CCD performance properties and special electrical circuits, a CCD camera is capable of preventing undesired highlights from spoiling images.

- *Operating life*—CCDs do not wear out and they are not easily damaged by pointing the camera at extreme highlights or even the sun. Therefore, they offer long life with consistent performance.

- *Maintenance costs*—As a result of the better stability and the long operating life, the life-cycle cost for a CCD camera is greatly reduced.

These points are further explained in the rest of this chapter.

3.2.1 CCD History

As solid-state integrated circuits developed, the dream of a solid-state imager appeared. It took a long time to be realized in the form of charge-coupled devices. A CCD is a repetitive, two-dimensional, array of cells that performs photodetection, storage, and scanning. The array is fabricated by solid-state processing techniques such as deposition, diffusion, and etching, which produce a monolithic unit of very high mechanical and electronic precision. In fact, the requirements for a CCD array are so demanding that it has taken

---

* There are other solid-state imager technologies, such as MOS (metal-oxide-silicon) devices, but the CCD is by far the most common type.

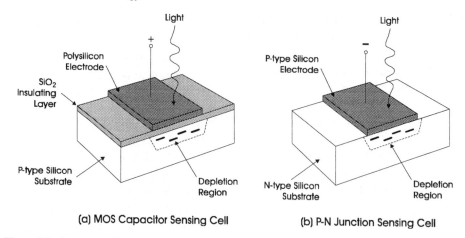

Figure 3.1 *Operation of (a) MOS capacitor and (b) P-N diode sensing and storage cells.*

more than 20 years since the first development of CCDs until their widespread deployment in all types of video cameras.

CCD arrays are different from most integrated circuits because their signal handling is analog. Minute variations in processing or local defects in the materials can cause image impairments that make the device unusable. Thus, they require different manufacturing process concerns than digital ICs to achieve high quality. For many years, this limited the use of CCDs to low-quality applications, and high-quality CCDs were very expensive. Now, those problems have been overcome, and CCDs are competitive over the full range of camera applications from low-cost cameras to the highest-performance cameras.

3.2.2 CCD Operation

The operation of a CCD may be understood by considering the structure of a single cell. Figure 3.1(a) shows the basic principle in a rudimentary cell consisting of a single capacitor constructed from a semiconducting material. A silicon-dioxide (SiO_2) insulating layer is deposited on a P-type silicon substrate. A conducting electrode above the SiO_2 layer provides one capacitor plate and the substrate is the other plate. This is called a *metal-oxide-semiconductor* (MOS) structure and this particular device is an MOS capacitor.

When a positive voltage is placed on the conducting electrode relative to the substrate, a *depletion region* is formed in the substrate beneath the electrode. This is simply a region of the substrate that has an affinity for free electrons. A depletion region is sometimes called a *potential well* because of its appearance in a diagram of the surface potential across the structure. Electrons "fall" into the well and are trapped.

If light impinges on a depletion region, free electrons (and holes) are formed. The holes are recombined within the substrate, but the electrons remain trapped in the depletion region. Thus, a negative charge builds up in the region that is proportional to the integrated light over time. To make it possible for light to reach the depletion region, the conducting electrode is made from *polysilicon*, which is a transparent, highly conducting form of silicon.

Another architecture that behaves the same way with light is a reverse-biased P-N junction diode. As shown in Figure 3.1(b), a depletion region also exists in the substrate beneath the diode, and light reaching this region will create free electrons that remain trapped (stored) there.

Either the diode structure or the MOS capacitor perform the functions of photodetection and charge storage for an imager, but the most interesting feature of CCDs is how scanning readout is achieved. This is the "charge-coupling" and is shown in Figure 3.2. Charge in one depletion region will move to an adjacent region if the latter region has a higher voltage applied to its electrode. That makes the adjacent region "deeper" and the charge moves to the deeper region. By performing this voltage-changing operation sequentially, charge can be made to move across an entire array of cells. The signals that do this are called the *transfer clocks*. By applying multiphase transfer clock signals to an entire array, all the cells of the array are transferred one step for each complete clock cycle.

Figure 3.2(a) shows two complete MOS capacitor imaging cells adapted for three-phase transfer clocks. There are three separate electrodes for each cell, with one transfer clock phase connected to each of them. The clock waveforms are shown in Figure 3.2(b). During the charge-integration period, which is the period for build-up of the stored charge image, one of the clocks ($\phi2$) is held high and the other two clocks are low. Thus, there is a depletion region under the electrode connected to $\phi2$, and free electrons created by the optical input are collected in that region. This is shown in the first line of Figure 3.2(c). Note that this region under the $\phi2$ electrode will collect the charge created under all three electrodes of that pixel. The holes (+) created are recombined in the substrate.

When transfer begins, the $\phi3$ clock is made high and the $\phi2$ clock voltage is slowly reduced, as shown in the waveform diagrams of Figure 3.2(b). The result, shown in the second line of Figure 3.2(c), is that the charge under $\phi2$ moves to be under $\phi3$. At the end of the first part of the transfer, $\phi2$ reaches zero, and the charge is now fully under the $\phi3$ electrode. Transfers take place simultaneously in every pixel that is connected to the same three clocks. Subsequent parts of the clock cycle transfer the charge progressively to each electrode at the right. At the end of one full cycle, the charge from the first pixel is now under $\phi2$ of the second pixel.

With proper design of the structure and the driving waveforms, charge transfer can be made fast and efficient, allowing thousands of transfers with negligible loss of charge and transfer speeds up to the tens of MHz range. Although Figure 3.2 shows a three-phase structure, many CCDs use a two-phase design and some even have four-phase clocking.

3.2.2.1 Imaging Cell Features

The cells of a practical imager contain additional features to support the array architecture. Two of these are explained here.

- *Channel stops*—barriers are built into the CCD structure to constrain the charge from spreading or transferring in the wrong directions. For example, a portion of

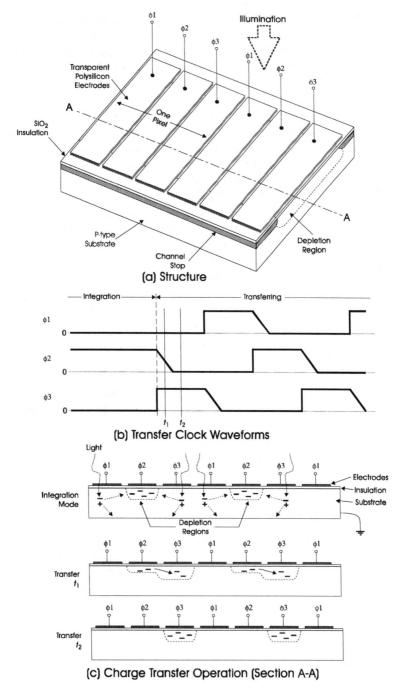

(a) Structure

(b) Transfer Clock Waveforms

(c) Charge Transfer Operation (Section A-A)

Figure 3.2 *Charge coupling.*

the array that transfers horizontally must have horizontal barriers to prevent the charge from moving vertically.

- *Overflow drains*—the formation of free electrons does not stop when a cell's well becomes filled. Therefore, the cell design must include a means to drain off the extra charge to prevent it from spreading into adjacent wells. That is the purpose of overflow drain structures.

Additional cell features are covered in the discussion about specific CCD architectures in the next section.

3.3 CCD ARCHITECTURES

There are three fundamental types of CCD array architectures. All three are used in different designs, sometimes with other modifications that deal with problems inherent to each architecture. These array architectures are discussed here.

A CCD array is designed to create a stored charge image of the entire scene simultaneously. Separate sensing cells are provided for each pixel of the entire image. Usually this takes the form of rows of cells, one row for each scanning line called for by the video system standard. (This may not be the case for interlaced scanning—see Section 3.4.3.) In each row, there are sufficient pixels to achieve the desired horizontal resolution—see Section 3.4.1. Note that the requirement to have one row of cells for each scanning line is the simplest case; with digital memory and processing in the camera, it is completely practical (and sometimes advantageous) to support a video system standard having a different number of lines than the number of rows of pixels in the imagers (see Section 4.9).

3.3.1 Frame-Transfer Architecture

The simplest and easiest-to-manufacture CCD architecture is the frame-transfer (FT) device. In this architecture, shown in Figure 3.3, there are two complete arrays of cells, each containing all the pixels of the image. Although this might mean that an FT chip is twice as large as the optical image format, it was considered a good approach in the early days of CCD arrays. (Note that a practical FT chip does not have to be twice as large as the image because the second array may be reduced in size vertically.)

One array receives the optical image and this is where integration occurs. The cells of this array are designed to shift vertically and, during the vertical blanking interval, the entire charge image is shifted as rapidly as possible down into a second array, which is masked from light by an opaque coating. This transfer process clears the pixels of the imaging array so a new integration period can begin.

During horizontal blanking intervals of the next vertical scan, this second array is further transferred down one line at a time into a *readout register*. During active horizontal line scan, the readout register is clocked to transfer the charge representing the pixels horizontally into an output gate, where they are converted into a video signal.

Note that the FT configuration shown in Figure 3.3 requires the optical image on the

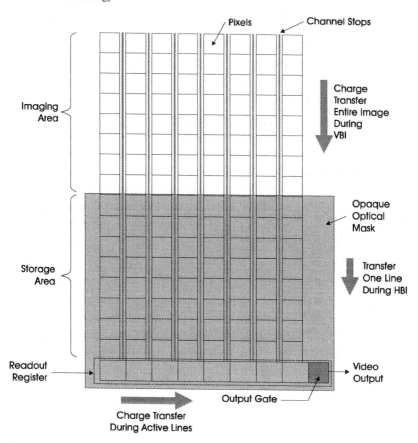

Figure 3.3 *Frame-transfer CCD architecture.*

imaging area to be rotated 180 deg as it is when imaged by a single lens. Thus, the lower-right pixel of the charge image is the first one clocked out of the readout register after the VBI. With the optical rotation by the lens, this corresponds to the upper-left of the scene.

Remembering that charge integration occurs continuously in the pixels as long as light is on the array, it can be seen that some small amount of integration will occur even during the VBI time when the vertical transfer is taking place. Since the charge image is moving during the transfer, the charge created during this time will be smeared vertically. The VBI represents about 6% of the total integration time and the smeared artifact may be larger than 6% by the amount that a highlight exceeds scene white level. It appears as a vertical line above and below the highlight. This is called *transfer smear* and it is a serious problem in FT imagers.

The only way to prevent transfer smear completely is to cut off the light during the vertical transfer. That can be done with a mechanical shutter in the camera and early FT cameras actually had such a device. However, a mechanical device is undesirable in an

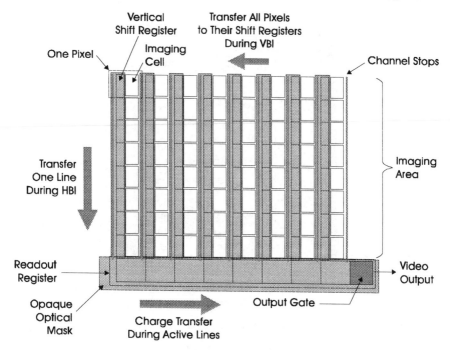

Figure 3.4 *Interline-transfer CCD architecture.*

otherwise all-electronic camera, and other CCD architectures were soon developed to avoid the transfer smear problem.

3.3.2 Interline-Transfer Architecture

By interleaving the imaging and storage arrays into a single structure called an *interline-transfer* (IT) CCD, the transfer smear problem can be largely avoided (see Section 3.4.10.1). As shown in Figure 3.4, each pixel in an IT imager has two CCD cells side-by-side. One cell performs integration and the second cell, which is masked from light, is the vertical-transfer shift register equivalent to the bottom array in the FT structure of Figure 3.3. During VBI, all the charge collected in the integration cells of all pixels is shifted horizontally into the adjacent vertical shift register cells in one massive single transfer. Then, the vertical shift registers are clocked down one line at a time and read out the same as in an FT device.

Since the optically-masked vertical shift registers are within the image area, there is a loss of light energy because light falling on the optical masks is not converted to charge and integrated in the imaging cells. This causes a 50% or greater loss of sensitivity. For most applications, that is not serious because CCDs have more than enough sensitivity. Reducing the width of the integration area also causes an improvement in MTF (see Section 3.4.11). Sensitivity can be restored for the most demanding applications by using

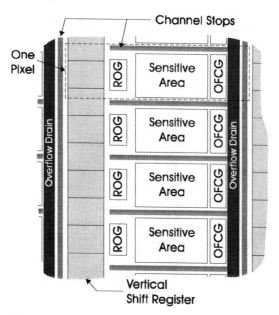

Figure 3.5 *Enlarged view of an IT imager showing the structure of individual pixels.*

a technique of placing a miniature lens over each pixel on the chip to direct all of the incident light onto the imaging cell. This is called the *on-chip lens* technique and is used in the highest-performance IT devices.

Since the horizontal transfer to the vertical shift registers has only a single step, there is no possibility of smearing while that occurs. All vertical transfers take place in the optically masked registers and smear cannot occur there either. Thus, transfer smear is virtually eliminated. However, a similarly appearing artifact may still occur on extreme highlights, although at a much lower level than in an FT imager. This has several causes: some light may leak around the optical masking of the vertical shift registers; or long-wavelength (red) light, which penetrates deeply into the substrate, may generate charge that migrates into the vertical shift register. This artifact is called *vertical smear,* and it produces a vertical line above and below highlights that looks much the same as transfer smear in an FT imager, but it is much lower in level.

The actual structure of a pixel is more complex than implied by Figure 3.4. Figure 3.5 shows more detail of the pixels in an IT imager. The region marked "Sensitive Area" is the only part of a pixel exposed to light; the rest of the pixel is optically masked. Channel stops are required both horizontally and vertically as shown. The readout gate (ROG) operates at the appropriate time allowing the stored charge to move from the sensitive area into the vertical shift register. In the case of an extreme highlight where the storage well of the sensitive area becomes full, the overflow control gate (OFCG) allows excess charge to flow to the overflow drain, where it goes to ground instead of damaging adjacent pixels.

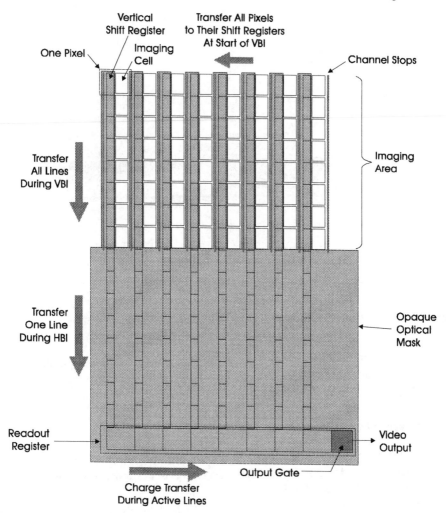

Figure 3.6 Frame-interline transfer CCD architecture.

3.3.3 Frame-Interline Transfer Architecture

The small transfer smear problem of the IT imager can be further reduced by the *frame-interline transfer* (FIT) CCD structure, which combines the concepts of the FT and the IT architectures. This is shown in Figure 3.6. This may be the ultimate CCD architecture developed so far, and it is used in the highest-performing cameras.

The operation is the same as the IT imager during the integration period, but all pixels are moved to their vertical shift registers at the start of the vertical blanking interval, following which they are immediately shifted down into the optically masked storage register. These transfers typically happen at 60 times the line-scanning rate, which means

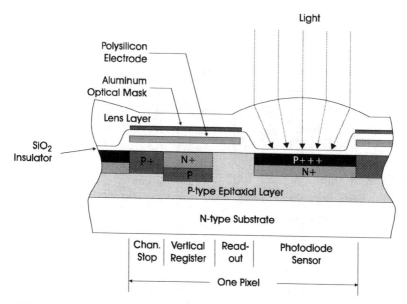

Figure 3.7 *Cross section of one cell of a Sony Hyper HAD imager. (Courtesy of Sony.)*

that vertical smear is reduced by the same factor. That makes the effect negligible. During active scanning time, lines are shifted into the readout register the same way they are with an FT imager.

3.3.4 Advanced CCD Designs

Each CCD manufacturer has developed proprietary cell design technology to enhance the capabilities of their devices. One of these that incorporates most of the features is the Hyper HAD™ technology developed by Sony [2].

The hole accumulation diode (HAD) device is a photodiode type of sensor built on a special substrate to improve sensitivity, highlight handling, resolution, and noise performance. The Hyper HAD device additionally includes on-chip lenses to further improve sensitivity. A cross-section of a Hyper HAD cell is shown in Figure 3.7. The substrate is N-type that has a P-type epitaxial layer grown on it. A heavily doped P region (P+++) functions as a hole accumulation layer. It is connected to the channel stop and serves to draw away the holes produced by photons. The combination of this region and photodiode operation gives the name hole accumulation diode.

The N-type substrate allows overflow charge to be drawn off into the substrate, which eliminates the need for the overflow drains that use up chip area in other designs. Thus, the percentage sensitive area of the pixel can be increased and overall pixel dimensions can be smaller. This allows higher resolution in the same chip size. The other function of the N-type substrate is to prevent free electrons generated deep within the substrate from reaching the storage wells—they are trapped in the substrate. This prevents the vertical smear effect.

The figure also shows the microlens layer that directs most of the light incident on a pixel to the sensitive area.

3.4 CCD PERFORMANCE

In many respects, performance of the CCDs will determine a camera's overall performance. This section discusses CCD performance factors and how they affect camera performance.

3.4.1 CCD Resolution

To a first approximation, the number of pixels in the CCD limit the ultimate resolution of a camera. One cannot obtain more resolution from a camera than the capability of its CCDs, but there are other factors that may reduce the camera's resolution including the camera's lens and other optical components.

Because of the discrete pixel structure, a CCD is spatially sampled in both the horizontal and vertical directions. However, the signal level of each pixel is an analog quantity, so quantizing is required to make a fully digital signal. But the resolution is limited by the sampling, which is discussed in the following sections.

3.4.1.1 Modulation Transfer Function

Many camera specifications give a single number for the camera resolution, which is the *limiting resolution*. That is the highest resolution that is observable in the output of the camera when viewing a pattern of closely spaced white and black lines, and it is usually given in TVL. A much better specification of resolution is the *modulation transfer function* (MTF), which is the curve of signal output level versus the spatial frequency of a pattern of alternate black and white lines. A system's MTF curve is the most complete specification of resolution performance, but cameras seldom specify this completely. At best, a single point on the MTF curve may be specified.

The optical input to a CCD is integrated over the sensitive area of each pixel. This is equivalent to sampling an electrical signal with a wide pulse, which causes a reduction of output at high signal frequencies when compared to sampling with narrow pulses. Thus, there is an MTF loss associated with sampling with pixels of significant width relative to their spacing. This is a $\sin(x)/x$ function of frequency that is governed by the percentage of a pixel's width that is sensitive, as shown in Figure 3.8. For full-width pixels ($W = 100$) the MTF is reduced to 63.6%. This loss is usually corrected in the camera signal processing circuits.

IT and FIT imagers usually have sampling widths less than 50 in the horizontal direction because of the presence of the optically masked bertical transfer registers next to

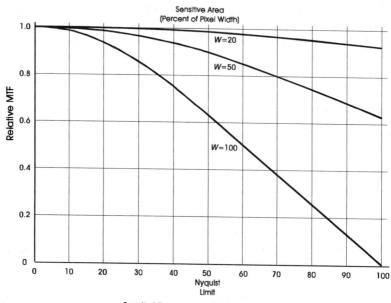

Figure 3.8   *Effect of sampling width on MTF.*

each pixel. This provides improvement in horizontal MTF at the expense of sensitivity. However, if on-chip lenses are used, the sampling width becomes as wide as the light collection area of the lenses.

Section 1.5.2.1 discussed the Nyquist limit that applies to all sampling processes—the sampling frequency must be at least twice the highest signal frequency to be reproduced without aliasing. To avoid aliasing, a prefilter is required before sampling to remove any frequency components in the input that are higher than the Nyquist limit. Since a CCD samples an optical image, the sampling prefilter must be in the optical path. Special optical low-pass filters (LPF) have been designed for this purpose and a high-quality camera contains such a component in its optical path.

3.4.1.2 How Much Resolution?

Resolution, especially in the horizontal direction, is a very competitive issue among different cameras and has been creeping steadily higher in all categories of cameras. Cameras always specify a limiting resolution number, but this may have no relationship to the resolution capability of the system using the camera. Unfortunately, limiting resolution is often a major item of camera competition, but the MTF curve, which is often ignored in camera specifications, is a much more important factor. Resolution adds significantly to cost and one really shouldn't invest in more resolution than the application requires. There are several matters to consider when deciding how much resolution is needed:

1. The capability of the final medium that delivers the video to viewers. If a camera is delivering signals that only will be broadcast over an NTSC channel with a bandwidth of 4.2 MHz, the viewer will never see more than 340 TVL resolution regardless of how high the camera resolution. However, it is desirable to maintain a high MTF out to the channel limit, which means that the camera limiting resolution number should be somewhat higher than 340 TVL. On the other hand, if the same video may be sent directly to monitors over high-bandwidth channels, viewers will be able to see higher camera resolutions.

2. The capability of video recorders in the system. Generally, the resolution limit of recorders is less than contemporary cameras, even for digital recorders. Like the distribution medium, recorders usually have a sharp resolution cutoff and the camera objective should be to maintain high MTF out to the recorder cutoff. Resolution beyond that point is immaterial, as no one will see it after recording.

3. The amount of processing of the signal that will be done along the way. Various special effects, image enhancement, or video compression processes may benefit from higher resolution. When these actions are expected to be needed, resolution should not be short changed in cameras or recorders. A high resolution in the original signal capture provides the maximum flexibility for the future uses of the video.

3.4.1.3 Horizontal Resolution

The number of pixels across the imager determine the horizontal resolution. Since TVL is defined as the number of both white and black lines that are contained in a distance equal to the picture height, the limit of horizontal resolution is the horizontal pixel count divided by the aspect ratio. Thus, a 1000-pixel CCD for a 4:3 aspect ratio has a theoretical maximum horizontal resolution of 750 lines. That number also is the Nyquist limit for the device and represents the highest possible resolution of the imager. In practice, the MTF at that resolution will be significantly down because of sampling width loss and the optical prefilter. An MTF of 50% at a frequency close to the Nyquist limit is very good performance. Since the system response must be essentially zero above the Nyquist limit to avoid aliasing, it is not practical to have too-high MTF response close to the Nyquist limit because that would require too sharp an optical filter cutoff, which is difficult to realize.

The number of horizontal pixels is a trade-off between CCD chip size, sensitivity, and manufacturing cost. Since the signal output from a pixel at a given illumination will depend on the sensitive area of the pixel, more pixels either means smaller pixels in the same size chip (less sensitivity and higher chip cost) or a larger chip size to accommodate more pixels of the same size. A larger chip costs more and also means larger optics, which is higher in size, weight, and cost. Thus, one can expect that higher horizontal resolution will always cost more, everything else being equal.

Most cameras operate their digital processing at the horizontal readout transfer clock frequency of the CCD, which means there is one digital pixel for each CCD pixel.

3.4.1.4 Vertical Resolution

In the vertical direction, the line-scanning pattern is an inherent sampling process that cannot be avoided. It is usually advantageous to choose the CCD vertical pixel number to match the active line number. That avoids introducing another, different, sampling process.

The only case where this is not true is when an HDTV camera is used to deliver a 525- or 625-line output. Then, the CCD vertical sampling rate is significantly higher than the video system and vertical scan conversion is used to reduce the vertical rate. With digital processing and filtering, an excellent job can be done; in fact, the 525- or 625-line signals produced this way can be better than are possible with a 525- or 625-line camera. That is because the higher initial line sampling followed by downconversion and filtering allows much of the vertical alias artifacts to be eliminated and a higher in-band MTF profile to be realized.

The theoretical maximum vertical resolution of a video system is equal to the number of active lines and this is also the Nyquist limit for the system. With appropriate optical prefiltering, that resolution can be approached in a CCD camera.

Many video systems use interlaced scanning, which requires special considerations in CCD imagers and affects the vertical resolution. This subject is discussed in Section 3.4.4.

3.4.2 Sensitivity

High sensitivity is desirable to allow shooting under low-light conditions and to operate at higher light levels with the lens stopped down for good depth of field. It is usual to specify camera sensitivity at two light levels:

1. Light level required for full-quality pictures at a specified lens f-number, scene illumination, and maximum scene reflectance. Typical performance is 2000 lux at f/8 with 3200K illumination and 89.9% scene reflectance. This is called the "sensitivity" specification and the same conditions are used for the SNR specification.
2. Light level required for "high-gain" operation (see Section 4.3), with the lens wide open. Typical performance is 2 lux with an f/1.4 lens and +30 dB gain. This is called the "minimum illumination" specification.

The sensitivity performance of a CCD depends on the percentage of a pixel's area that is sensitive, the quantum efficiency in converting photons to hole-electron pairs, the optical efficiency of light reaching the sensitive area, and any losses in retrieving the stored charge. Of course, the specifications above also depend on the camera's optical and electrical efficiency.

There is no sensitivity advantage or disadvantage to optical format sizes. Because a larger format requires longer focal length lenses; at the same f-number and number of pixels, all format sizes have the same sensitivity. However, within a given format size, sensitivity reduces as the resolution is increased. That is because pixels become smaller and the amount of light falling on each pixel is reduced.

3.4.3 Signal-to-Noise Ratio

Noise is an important factor in consideration of camera sensitivity. Theoretically, a camera's sensitivity could be increased as much as desired simply by increasing amplifier gain and operating the CCDs at a lower output level. Of course, the signal-to-noise ratio (SNR) will degrade when this is done, approximately by a factor equal to the amount that the gain is increased. That is what happens in a camera's "high gain" modes, which trade signal quality for sensitivity.

The full-quality mode of a camera operates the CCDs at the light level given in the sensitivity specification. As explained in Section 3.4.8, this may be somewhat of a trade-off with highlight performance, but most cameras deliver a full-performance SNR in the range between 50 and 60 dB. These numbers produce subjectively noise-free pictures. However, the higher numbers allow more postprocessing of the picture to enhance sharpness or other picture attributes.

Noise generation in CCD imagers has several sources. The fundamental noise level results from the quantum nature of the incident light—as the light on a pixel reduces, fewer light quanta and, thus, fewer electrons are involved, and the signal gets noisier. However, this is usually not a serious limit. More important is the dark current performance of the CCD; this is a small current that flows in the absence of light input. It depends on temperature and may vary from pixel to pixel. Random fluctuations of dark current are visible as random noise, but the pixel-to-pixel variations will appear as a pattern if they become great enough. This is *fixed pattern noise* and it directly depends on the quality of the manufacturing process for the CCD. Another CCD noise source is reset noise, which originates in the readout circuit on the chip. That can be controlled by appropriate design of the off-chip readout processing (see Section 4.3.1). The input amplifier is also a noise source, but CCD output levels are high enough that amplifier noise is usually negligable in the total except possibly in the highest gain mode.

3.4.4 Interlaced Scanning

As explained in Section 1.6.3, existing TV systems use interlaced scanning to conserve bandwidth while maintaining low flicker in displays. Interlacing requires that alternate lines of the raster be scanned on one vertical scan, with the lines skipped being scanned on the next vertical scan. It requires special consideration in CCD imagers.

3.4.4.1 Interlacing in FT Imagers

In FT CCDs, all pixels must inherently be shifted out for every vertical scan. One approach for interlacing an FT imager is to have only half of the total lines in the CCD array, and arrange that the spatial integration location of a pixel shifts vertically by one line

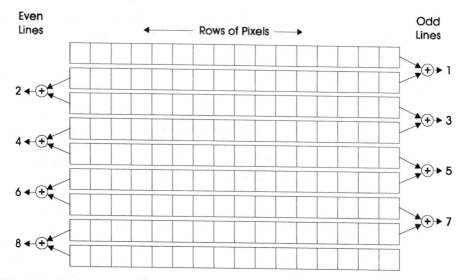

Figure 3.9 *Interlacing of an IT or FIT CCD by combining adjacent lines.*

between alternate vertical scans. That is easily done in a two-phase or four-phase array by changing the segment of the pixel that is held high during integration. This shifts the vertical location of the depletion region collecting the electrons and effectively moves the spatial location of the pixel. On readout, all the lines of the array are read out for each vertical scan, but they represent two different spatial locations.

This approach gives good interlacing, but it causes a loss of vertical resolution because each pixel is effectively two lines high. Since the sensitive region of a pixel is nearly full height vertically, this is equivalent to sampling with a pulse that is twice as wide as one line ($W = 200$, see Section 3.4.1.1), and the sampling width loss is severe. The MTF actually goes to zero at a resolution value equal to the line number.

Because all pixels are shifted out on every field scan, light is integrated over only one field period. This is called *field integration*.

3.4.4.2 Interlacing in IT and FIT Imagers

In IT and FIT imagers, separate pixels are provided for odd and even lines and the readout is determined by controlling which pixels are transferred to the vertical shift registers during vertical blanking. That provides full vertical resolution, but it means that light is integrated for a full frame in each pixel. This is *frame integration*, which gives greater motion blurring than field integration. (It also gives higher sensitivity because of the longer integration time.)

Interlacing can also be achieved in IT or FIT imagers by combining the outputs of vertically adjacent pixels on each field to achieve the motion performance of field integration without a loss of sensitivity. This gives less blurring on motion, but it also causes

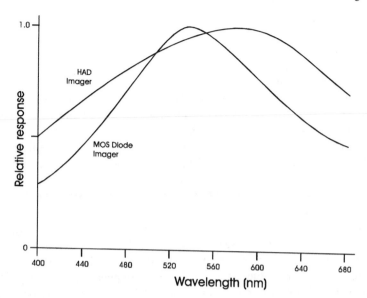

Figure 3.10 Spectral response of typical CCDs.

reduced vertical resolution for the same reason explained for FT imagers. However, this approach is considered a good compromise and most imagers use it because the digital filtering inherent in the combining of pixels vertically allows reduction of vertical aliasing and a better MTF curve. This readout architecture is shown in Figure 3.9. Vertically adjacent lines are combined during readout of odd and even fields. The result is perfect interlace with a vertical aperture height of two pixels. Because the outputs of two pixels are combined, the output signal is doubled, giving 2× higher sensitivity compared with progressive scanning of the same CCD.

3.4.5 Spectral Response

As explained in Section 2.2, a video camera must have specific spectral responses in each of its color channels (usually red, green, and blue) according to the standards for the video system. This is accomplished by the combination of the spectral responses of the CCDs, the light-splitting optics, and any additional "trim" filtering. The response achieved this way can be further modified by electronic signal processing (masking) to meet the system requirements.

Since all the CCDs in a camera are usually the same type, the CCD ought to have a broad spectral response to cover all colors. Most CCDs have good red (long wavelength) response, but blue response can be a problem because of absorption in the polysilicon layer that covers the sensitive area. Note that the HAD imagers do not have polysilicon over the sensitive area, so they have better blue response. Figure 3.10 shows this.

3.4.6 Spatial Offset

CCD resolution costs money and techniques for achieving more resolution without increasing the CCD resolution are very desirable. In a three-CCD camera, it is possible to achieve an effective horizontal resolution for the luminance channel corresponding to nearly twice the number of horizontal pixels in each CCD. That is done by precisely offsetting the CCDs by one-half a pixel width in the horizontal direction. When signals are combined for the luminance channel, the sampling rate is effectively doubled.

If the red and blue CCDs are shifted relative to the green CCD by one-half of the pixel width, and the R + B signals are combined equally with the G signal, there is cancellation of aliasing components up to the sampling frequency of each CCD, effectively doubling the Nyquist limit. However, the luminance channel of most systems does not combine those components equally (for NTSC or PAL, R + B is 0.41 and G is 0.59), so the cancellation is not complete. It is still an improvement, but better results can be achieved by using two green CCDs and two more for red and blue. This requires four CCDs and more complex optics, but it delivers the most resolution. It is no longer used because good enough results now can be achieved with three-CCD architectures.

An alternative approach is to have two green CCDs but only one CCD to reproduce red and blue. Combining red and blue is done with the same techniques used for single-CCD cameras (see Section 3.6). This results in lower resolution in those color channels, but most systems use color subsampling anyway, and it may be an acceptable trade-off for some applications.

Spatial offset is more effective in IT or FIT imagers than it is in FT imagers because the latter devices have full-width sensitivity of the pixels, so the MTF falls to zero at the CCD sampling frequency. Although the spatial offset can cancel aliasing up to this frequency, it does nothing for the MTF loss caused by the width. However, IT and FIT imagers have a narrower sensing width to begin with, so their MTF curves are inherently better (see Section 3.4.1.1). Figure 3.11 shows how combining the outputs of two IT imagers with one-half pixel offset doubles the horizontal resolution.

Another consideration of RGB-style spatial offset is that its effectiveness may be impaired by lateral chromatic abberation in the camera lens. That can shift the red and blue images with respect to the green image, thereby interfering with the precise offset. This problem does not occur in the two-green types of offset, since both green images will see the same aberration from the lens. There have been some efforts to standardize lens chromatic aberration to minimize differences between lenses, thus making interchange of lenses easier.

Because of the significant improvement it offers, most three-CCD cameras use some form of spatial offset. A four-CCD configuration with two green imagers was used in some early HDTV cameras where the ultimate resolution was desired, but it is no longer in favor. Since CCD resolution numbers continue to improve and costs keep falling, this whole issue may some day become moot.

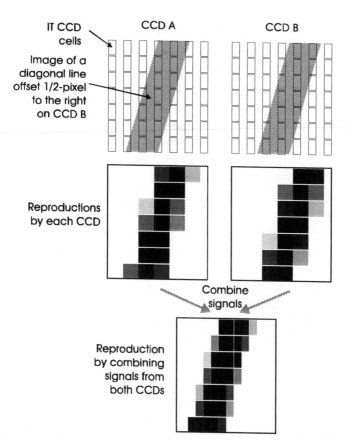

Figure 3.11 CCD spatial offset for improvement of luminance horizontal resolution. (Reproduced with permission from A. C. Luther, Principles of Digital Audio and Video, *Artech House, 1997.)*

3.4.7 Black Level Control

Proper reproduction of video signals requires that signals contain a reference black indication that can be set to reproduce black at the display. Reference black is generated in a CCD by providing pixels around the edges of the image that are masked from light. As shown in Figure 3.12, a typical CCD has more pixels than the final output will contain to provide this black reference. When the CCD is scanned, some or all of the black reference pixels are scanned to create a black region in the signal output. A method for establishing the signal's black reference by using the extra pixels is described in Section 4.2.1.1.

3.4.8 Shading

The output from a CCD is not necessarily uniform over the area of the device or from pixel to pixel. This nonuniformity, called *shading*, is an important quality measure. It is

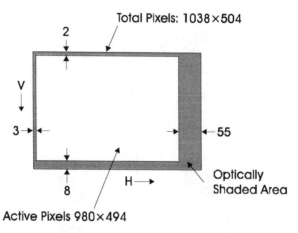

Figure 3.12 *CCD showing extra pixels for black level reference.*

usually characterized in terms of black shading and white shading. Black shading originates in the dark current of the CCD (see Section 3.4.3). White shading is not necessarily a CCD effect, but originates mostly from optical vignetting (see Section 2.3.3.2). Both shading effects can be electronically corrected, as described in Section 4.7.

3.4.9 Dynamic Range

The signal output from a CCD ranges from the noise level (called the *noise floor*) when there is no light; up to the maximum highlight output resulting from full-well conditions (called *sensor saturation*) in the CCD cells. In modern CCDs, this range can be 80 dB or more, which is far more than is necessary in any video system. Therefore, cameras are set up to operate their CCDs well below the saturation level for peak white video conditions. That means the CCD will deliver higher-than peak-white signals on highlights and this must be sensibly handled in the signal electronics.

High-quality cameras typically set peak white at 60 to 62 dB above the noise level, leaving as much as 20 dB of CCD dynamic range for highlight handling. The camera electronics are designed to either compress or clip signals above peak white. This gives excellent control of highlights up to 20 dB above white level. Beyond that, the overflow control built into the CCDs comes into play. That is inherently a hard clipping effect which cannot be controlled by the user, but it still prevents highlights from spreading or smearing. The availability of excess CCD dynamic range for controlled handling of highlights is a major advance in camera design (see Section 4.4). Nothing like it was possible with tube-type imagers.

3.4.10 Motion Effects

Tube-type imagers can have a problem because their scanning mechanism may not remove all the charge image on each scan; a certain percentage of charge can be left behind after a scan. That means each frame contains some information from previous frames, which causes a smearing effect on moving objects in the scene, called *lag*. FT CCDs inherently do not have lag, because the same charge transfer efficiency required to move the image through thousands of transfers for readout means that nothing can be left behind from one frame to the next. On the other hand, lag can exist in IT or FIT imagers because of poor transfer efficiency in the transfer from the sensor area to the vertical registers. This problem has been solved in the higher-quality broadcast CCDs and lag is now almost zero.

Section 3.4.4 explained that interlaced scanning of a CCD can provide a choice between field or frame integration. That is a trade-off between motion artifacts and vertical resolution that the user can make at any time in cameras equipped to offer the choice.

3.4.10.1 Vertical Smear

Because the optical input is applied to the surface of the CCD all the time, there can be a problem of spurious charge information getting into the vertical-transfer registers in any of the architectures. This is serious in FT imagers because the main transfer of the image from the integration array into the readout array takes a significant portion of the vertical blanking period, representing as much as 6% of the total time. As the image is transferring vertically, a small amount of charge will be generated in any pixels illuminated by a highlight. That adds to the signals transferring through, resulting in a vertical line that appears above and below any highlights in the image. It is so objectionable in FT imagers that a mechanical shutter is always used in FT cameras to remove the optical input during the vertical transfer. This completely eliminates the problem for FT imagers but it is an awkward device to put in an otherwise all-electronic camera.

IT and FIT imagers eliminate most of the smear problem by rapidly transferring the entire charge image horizontally into the vertical-transfer registers that are adjacent to each pixel. However, even though the vertical-transfer registers are optically shielded from the light image, there can still be a problem of small amounts of charge from the image leaking into the vertical registers during the readout. Different architectures address this problem to various degrees, but the problem is never totally eliminated. Some consideration of numbers will show how small the smear effect has to be before it is not visible under extreme shooting conditions.

The most difficult situation occurs in low-light shooting where the camera's video gain is increased for maximum sensitivity. For example, a nighttime street scene may require maximum-sensitivity operation; the smear effect becomes noticeable if an automobile comes into the scene with its headlights pointed directly at the camera. This is an extreme highlight situation where the headlight intensity may be 1,000 or more times the desired scene highlight brightness. The CCD's overflow drain mechanism and the electronic highlight clipping will take care of the highlight, but any of that extreme energy getting into the transfer registers will produce smear. Consider that a smear as small as 40 dB below reference white level is easily seen when it moves across the picture; the 1,000:1 highlight is +60 dB, so the smear must be down 100 dB from the highlight level to not be

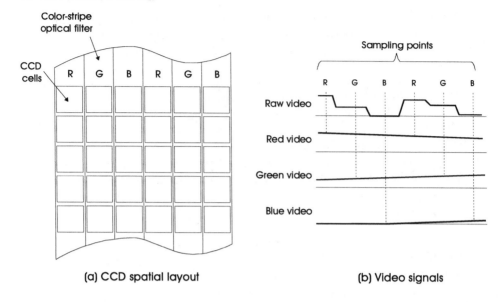

(a) CCD spatial layout   (b) Video signals

Figure 3.13 *Single-CCD color camera. (Reproduced with permission from A. C. Luther,* Principles of Digital Audio and Video, *Artech House, 1997.)*

seen at normal camera gain. Practical experience with extreme shooting conditions has shown that even this is not good enough for high gain operation and smear performance numbers as good as -145 dB are meaningful. IT imagers can deliver smear performance in the range of -105 to -120 dB; FIT imagers are in the range of -120 to -145 dB. For many users, this justifies the higher cost of FIT imagers.

3.4.10.2 Electronic Shuttering

Sometimes the full-time charge integration of the CCD's sensor cells is not desired because it represents an exposure time of a field or even a frame depending on the mode of operation. This causes blurring of moving objects, which is not a problem with motion video but it is important if the video is played in slow motion or still frame. Fortunately, most CCDs can be operated with shorter integration times, a technique called *electronic shuttering*. By controlling the application of voltages to the integration cells, this feature shortens integration time (at reduced sensitivity, of course) to reduce motion blurring.

Two other important features can also be achieved with electronic shuttering of FIT imagers. These are: (1) control of shutter rate to minimize the flicker effect that occurs when computer displays are present in the scene, and (2) extending vertical definition by shuttering of alternate rows of CCD cells to overcome the loss of vertical resolution described in Section 3.4.4.2.

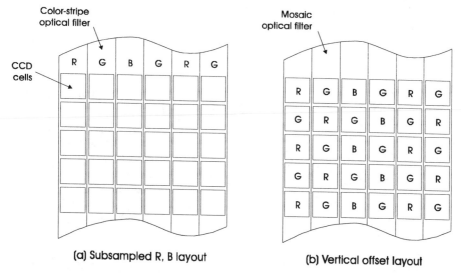

(a) Subsampled R, B layout   **(b) Vertical offset layout**

Figure 3.14 *Alternate color filter structures for single-CCD color cameras.*

3.5 SINGLE-IMAGER CAMERAS

Most of the discussion of imager applications so far has assumed there are three imagers and light-splitting optics in the camera. This approach dominates the professional and broadcast markets but it is too expensive for the consumer market. Single-imager color pickup has been developed for this market. Single-imager technology was developed with tube imagers and the same approaches are applicable to CCDs. A color filter structure is overlaid on the imager so that three adjacent pixels each respond to separate primary colors. When the CCD is scanned out, this produces a time-division multiplex signal as shown in Figure 3.13.

Figure 3.13(a) is a detail of the spatial layout of sensor cells on a CD. A vertical-stripe color filter is deposited on top of the CCD surface so that adjacent pixels see red, green, and blue light components. The output signal after scanning, shown in Figure 3.13(b), sequences between the pixel values for the three colors. By sampling this waveform three times corresponding to the timing of the color segments, R, G, and B component signals are retrieved. Of course, the color stripe filter must be precisely aligned with the CCD pixel structure; because of this, it must be applied during CCD manufacture, using the same kinds of processing that builds the CCD itself. Once that has been done, a monolithic color imager is obtained, and no further mechanical adjustment will be required (or is possible).

The luminance signal from an RGB stripe-filter color CCD has nearly a 3:1 loss of horizontal resolution compared to the CCD's total number of pixels. The same can be said for each of the RGB color components. This limits the use of stripe-filter color CCDs to applications that can tolerate some loss of resolution. However, some of the loss can be

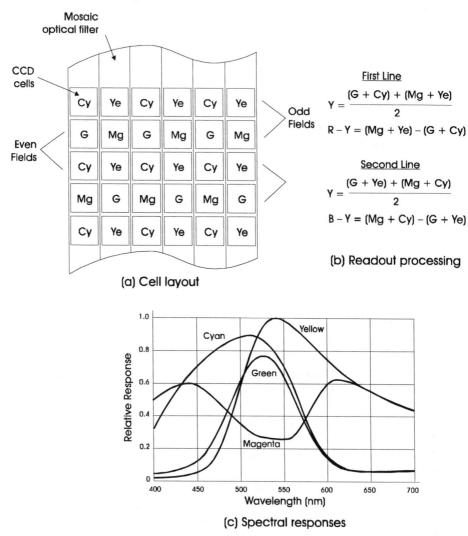

(a) Cell layout

(b) Readout processing

First Line

$$Y = \frac{(G + Cy) + (Mg + Ye)}{2}$$

$$R - Y = (Mg + Ye) - (G + Cy)$$

Second Line

$$Y = \frac{(G + Ye) + (Mg + Cy)}{2}$$

$$B - Y = (Mg + Cy) - (G + Ye)$$

(c) Spectral responses

Figure 3.15 *Single-device color CCD using complementary-color filtering. (a) cell layout, (b) readout calculations, (c) spectral responses.*

overcome by the use of more complex color filter structures. Recognizing that color-difference channels can have somewhat reduced bandwidths, the stripe filter can be revised as shown in Figure 3.14(a).

By having twice as many green cells as red and blue cells, the luminance resolution is improved to approximately a 2:1 loss compared to the total pixel count, while the color channels have 4:1 loss, corresponding to 2:1 subsampling of color difference signals relative to the luminance.

Table 3.1
Specific CCD Designs

Item	Sony BVP-Series Component CCD	Sony ICX068AK Color CCD	BTS Experimental HDTV
TV Standard	NTSC	NTSC	1125-30 HDTV
Aspect Ratio	4:3	4:3	16:9
Pixel Array (h × v)	1038 × 504	811 × 508	1920 × 1080
Total Active Pixels	480,000	380,000	2,070,000
Optical Image Size (in)	2/3	¼	1
Architecture	FIT	IT	FT
Construction	Hyper-HAD	HAD	
Features	Microlens	Complementary-Color Filters	High Resolution
	Electronic Shutter	Low Smear	
	Low Smear		

A further improvement is achieved by making the color filter in a mosaic pattern as shown in Figure 3.14(b). The sampling points are offset on adjacent lines, so there is now a luminance sampling point at every horizontal pixel position, but half of them are on adjacent lines. This allows the camera signal-processing circuits to trade vertical for horizontal resolution.

Another single-CCD design uses a complementary-color mosaic filter structure as shown in Figure 3.15(a). In this design, the color filters are green, yellow, cyan, and magenta in the pattern shown. An interlaced-scan readout combines information from adjacent lines to produce luminance and color-difference signals as shown in Figure 3.15(b).The luminance signal is obtained by adding two adjacent cell pairs and color differences are obtained by subtracting the cell pair values. Each readout line provides one color difference signal, the other color difference comes out on the next line. The subsequent signal processing must use one-line memories to produce both color difference signals for every line.

Because the luminance calculation has a green pixel in every horizontal position, horizontal resolution is increased compared with other types of single-CCD color imagers. Similarly, because the channel spectral responses are broader as shown in Figure 3.15(c), the optical efficiency and sensitivity is improved. However, the subtraction calculation of color-difference signals causes a loss of SNR for color, but if the CCD SNR is high enough to begin with that it is not a problem.

3.6 SPECIFIC DEVICES

Practical CCD designs use combinations of the techniques discussed above. This section compares several designs. Table 3.1 compares the major features of the units described here.

3.6.1 Broadcast Camera CCD

The Sony BVP-700/750 series cameras [3] are top-of-the line broadcast cameras for studio and field use in NTSC or PAL television systems. They have a three-CCD spatially offset configuration based on a 1038 × 508-pixel Hyper-HAD CCD (NTSC) or 1038 × 594 (PAL) using a 2/3-in optical format. The cameras have high sensitivity and a limiting horizontal resolution of 900 TVL. The MTF without any enhancement is 70% at 400 TVL. This CCD is also available in a wide-screen format for 16:9 aspect ratio operation.

SNR performance is equally impressive—63 dB at normal operating light levels. The cameras have many other operating and system configuration features, but the items listed here are the ones that depend on the CCD performance.

3.6.2 Consumer Camera CCD

The driving consideration in consumer cameras has to be cost. Competitiveness depends on delivering the highest performance at any particular price point. That leads to the single-CCD approach with the smallest possible image format. The Sony ICX068AK CCD is designed for this market [4]. Using a 1/4-in optical format, it still provides 768 × 494 active pixels for NTSC and, with the complementary-color filter concept, it provides good sensitivity and resolution for the consumer market.

3.6.3 HDTV CCDs

The HDTV maximum-resolution scanning format calls for 1920 × 1080 active pixels. CCDs are now available for that full resolution in 2/3 in format from Sony, Panasonic, Hitachi, Toshiba, Kodak, and others;, but the one described in Table 3.1 is an early experimental unit developed by BTS Television of Breda, The Netherlands and introduced in 1992 [5].

REFERENCES

1. Fisher, D. E., and M. J. Fisher, *TUBE: The Invention of Television*, Counterpoint, Washington, D. C., 1996.

2. Sony Corporation, *Imagery*, Sony Brochure BC-00379, Sony Business and Professional Group, 3 Paragon Drive, Montvale, NJ.

3. Sony Corporation, *BVP-700/750 Product Information Manual*, Sony Brochure BC-00448, Sony Business and Professional Products Group, 3 Paragon Drive, Montvale, NJ.

4. Sony Semiconductor Company of America, ICX068AK Data Sheet, http://www.sel.sony.com/semi/ccdarea.html.

5. SMPTE, *Implementing HDTV*, SMPTE, White Plains, NY, 1996, pp. 90-100.

4

Camera Signal Processing

Although the basic optical-to-electronic conversion is done by the CCDs in a camera, additional processing of the signals is required to optimize video performance and make signals consistent with the external system. The fundamental processing steps are listed here and briefly described; the rest of this chapter will discuss them in detail.

Until recently, all camera signal processing was done with analog circuits, usually with integrated circuits and some discrete components. Now, competitive digital techniques are available, and most new cameras are being introduced with digital processing because of all the advantages discussed in Section 1.5.1. This chapter mostly covers the digital approaches to each signal processing task, although analog circuits occasionally will be discussed by way of comparison.

4.1 STEPS OF CAMERA SIGNAL PROCESSING

All cameras must perform the following signal processing steps:

- Video amplification—while the output level from a CCD is high enough to be above system noise levels, it is still too low for most other processes. An analog video amplifier extracts the CCD signal and brings the level up to a normal level while rejecting noise that may be on the CCD output signal.
- Contrast compression—the dynamic range of scene illumination is often higher than the range handled by video displays. Contrast compression is an additional transfer characteristic nonlinearity that is inserted to reduce the signal dynamic range to what a display can handle.
- Aperture correction—the lens, optics, and the CCD sampling process introduce MTF losses (see Section 3.4.1.1), which can be corrected in the signal processing.
- Image enhancement—additional enhancement may be added to improve the sharpness or noise performance of images.
- Highlight protection—signal processing is used to compress highlights so they will not cause overload of subsequent video circuits.

- Color correction—masking may be required to compensate for colorimetric errors in the optics or to match colorimetry in multiple camera installations.
- Gamma correction—the transfer characteristic of a CCD is quite linear over an extremely wide range of light level, but most video system standards call for a nonlinear transfer characteristic that compensates for the display transfer characteristic. Gamma correction performs this task.
- Video compression—digital video systems often use compression to reduce the bit rate for storage or transmission. This may be done in the camera or later in the system.
- Encoding—the output signal must have the format called for by the external system. NTSC or PAL encoding or digital bit stream encoding are typical output formatting tasks. In the case of two-unit camera systems that have a camera cable (see Section 6.4.1.2), additional encoding is required to multiplex all signals on the camera cable.

4.1.1 Signal Processing Block Diagram

Figure 4.1 is a block diagram of signal processing for a typical 3-CCD camera using digital processing. Analog video amplifiers ahead of the ADCs bring the CCD output up to a level suitable for the conversion process. They may also perform part of the highlight compression task (see Section 4.2.1.3) or the white balancing (see Section 5.2). Beyond the converters, all processing is digital. A computer central processing unit (CPU) in the camera connects to all units through a data bus and provides control and storage of all setup parameters. This capability is discussed in Section 6.7.

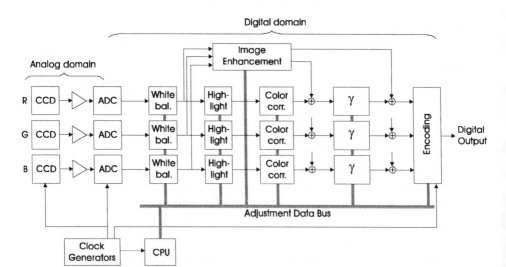

Figure 4.1 *Block diagram of typical digital camera signal processing.*

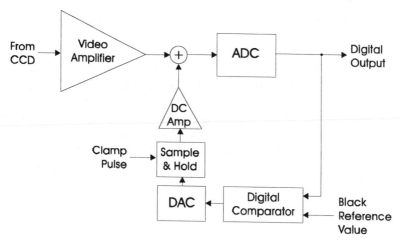

Figure 4.2 *Digital feedback clamp circuit.*

4.2 DIGITAL SIGNAL PROCESSING

There are a few generic digital processing techniques that may be used to accomplish most of the camera signal processing tasks. These are discussed in this section.

4.2.1 Analog-to-Digital Conversion

ADC was introduced in Section 1.5.2. However, there are several important considerations when converting the video signals from CCDs.

4.2.1.1 Black Level Control

A digital video system generally assigns a specific numerical value to picture black level. At the input of the ADC, the black level of the CCD analog signal must be held at the point on the ADC transfer characteristic that will deliver this specific digital value. In analog circuits, the process of holding black level to a specific value is called *clamping*, and that word is used for the same process in digital systems.

Clamping can be done by feedback around the ADC as shown in Figure 4.2. The digital value at the output of the ADC is compared to the desired black reference level. The output of the comparator, which is simply the digital difference between the inputs of the comparator, is converted to an analog signal by a DAC. This signal is sampled and held at the time of a clamp pulse, which is a pulse that identifies the time when the CCD is reading pixels that are shielded from light (see Section 3.4.7). The output of the sample & hold is a dc voltage, which is amplified and fed back to shift the dc level of the analog signal at the input of the ADC. The loop stabilizes with signal black at the desired digital value. Clamping must be done separately in each color channel of a three-CCD camera.

4.2.1.2 White Level Strategy

As with black level, a digital system specifies a numerical value for reference white level. This is usually substantially below the maximum digital value determined by the number of bits per sample, to provide overhead for highlights. A typical setting for a high quality camera is to set the CCD output saturation level (full-well) to near the maximum digital value and then operate 100% video at one-sixth of the full-well value. That allows a factor of 6 overhead for highlights. This subject is discussed further in Section 4.8.

The desired white level condition is established by setting the video amplifier gain. That may be done in the analog domain or it may be achieved by adjustments in the digital domain. The latter approach uses up some of the digital dynamic range, which may not be desirable.

4.2.1.3 Bits per Sample

The white level strategy and the camera SNR specification determine the bits/sample (bps) required for the ADC. For example, if the camera SNR is to be 60 dB and there will be 6× digital overhead for highlights, 12 bps conversion is required if the ADC quantization is not to limit the system performance. This can be calculated as follows:

> Quantizing noise should be at least 6 dB below the CCD SNR for it to not affect the resulting SNR. 6× is 15.6 dB so, adding 60 dB for the target SNR, 15.6 dB for the highlight overhead, and 6 dB for the quantizing overhead, we get 81.6 dB. Using (1.1) to calculate quantizing noise, we find that the closest match is 12 bps, which gives about 83 dB.

Many cameras do not use 12-bit ADCs, which means that some of the highlight compression should be done in the analog domain ahead of ADC. A common strategy is to perform analog compression of highlights down to 200% of reference white. This allows the ADCs to be 10 bps.

Once the highlight overhead has been removed by the highlight compression processing, the bps could be reduced to 10 bits or even 8 bits. However, there are other considerations in the processing that lead to the bps having higher values at different points in the processing (see Section 4.2.4).

4.2.1.4 Encoding

Most RGB digital cameras use simple binary encoding (see Section 1.5.2.3) in their ADCs. That is sufficient for the types of arithmetic required in camera signal processing.

4.2.2 Memory Look-up Tables

A digital system has a limited number of possible signal levels 2^N according to its bps number, N—8 bps = 256 levels, 10 bps = 1024 levels, 12 bps = 4096 levels, and so on.

Any amplitude transfer characteristic can be produced by calculating the output values for all possible input values and storing them in a memory. The transfer characteristic is applied to a signal simply by using the input signal value to address the memory and reading the addressed memory location for the output value. This is called a *memory look-up table* and it is widely used for transfer characteristic manipulation in digital cameras.

A 10-bit memory look-up table requires a 1024-entry memory with 10 bits per entry, which is a total of only 10,240 bits—not much memory at all; a 12-bit table requires only 49,152 bits of memory.

Since transfer characteristic processes such as gain control, gamma correction, or highlight compression typically require adjustment during operation, memory look-up tables usually use random access memory (RAM) memory circuits and are provided with an interface to a computer data bus in the camera for reloading of values. An embedded microprocessor in the camera performs calculations of new values and loads the memory tables during vertical blanking intervals. Thus, this approach can provide almost instantaneous adjustment of transfer characteristics. The microprocessor also allows multiple nonlinearities to be combined into a single memory table. The microprocessor performs the calculation of memory tabel values for the combined functions whenever any one of them is changed.

A look-up table can also perform linear multiplication by a constant. The table simply stores the input sample values multiplied by the constant multiplier. However, if multiplication of two signal values is required, an actual digital multiplier must be used. This is a more complex process but it is available in integrated circuits and is not expensive.

4.2.3 Digital Filters

Many camera processes, such as aperture correction or image enhancement, require modification of the frequency response characteristic of the digital signal. That is done with a *digital filter*, which is based on the frequency-domain properties of digital samples [1]. A digital filter simply sums weighted values of adjacent samples to produce any desired frequency response. This simple approach is called a *finite impulse response* (FIR) digital filter.

The weighting factor multiplications for digital filters can be done with look-up tables, but there may be a large number required (some filters use 21 or more sections.) Packaged digital filters generally use digital multipliers because of their greater flexibility. With multipliers, only a single value need be set per stage when coefficients are changed, whereas look-up tables require 2^N values per stage to be set.

4.2.4 Control of Bits per Sample

As signals are processed, the bits per sample (bps) have to be increased to preserve information generated in the procesing. For example, multiplying two 8-bit values yields a 16-bit value. However, the system output may support only 8 bps, so the 16-bit value has to

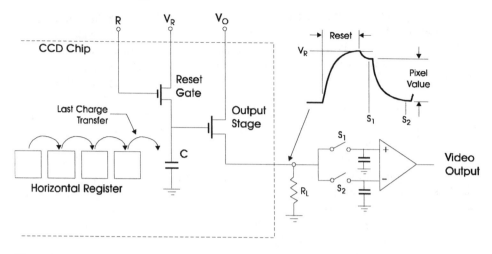

Figure 4.3 *CCD output amplifier and correlated double sampling.*

be rounded off sometime before the output. That is done by the process of *requantization*[2], which is a technique to prevent the introduction of new quantizing errors when rounding off unnecessary bits. However, repeated rounding can still cause accumulation of errors, and the best practice is to maintain a larger number of bits throughout the processing, rounding off only at the end.

The same consideration of extra bits comes into play with nonlinear processing. Section 4.7.2 discusses this matter with respect to gamma correction.

4.3 VIDEO AMPLIFIERS

The one unavoidable analog process in a digital camera is the amplifier that recovers the signal output from a CCD and brings it up to a level suitable for ADC.

4.3.1 Correlated Double Sampling

The output signal from a CCD is not a normal video signal because of the way that the charge transferred in the horizontal register has to be converted to a signal voltage for output. At the end of the horizontal shift register, each pixel's charge packet Q is transferred to a capacitor C, where it becomes a voltage $V = Q/C$ that drives an on-chip output amplifier, shown in Figure 4.3. However, since there is no further charge transfer following this capacitor, the charge from each pixel must deliberately be removed before transferring the next pixel. That is done by a gate to reset the capacitor to a fixed level before transferring each new pixel.

The result is that the output video signal is actually a pulse waveform at the pixel clock rate. The amplitude of each pulse represents the voltage of the corresponding pixel. Normally, this waveform passes through a *sample and hold circuit* to convert it to a video signal that changes smoothly from one pixel value to the next. However, there can be noise introduced to the signal by the reset process, crosstalk from the gating pulses, and by the on-chip amplifier that follows the output capacitor. This noise can be removed by taking two samples of the waveform and subtracting them to obtain the pixel value. This is called *correlated double sampling*.

Figure 4.3 shows the waveform resulting from the reset process. Sampling is done at point S_1 right after reset and then again at S_2 near the end of the charge transfer. The two sample values are held in capacitors and subtracted in the video amplifier as shown to obtain the output video without the reset noise.

4.3.2 High-Sensitivity Operation

All cameras require gain control of the input video amplifiers. Consumer cameras provide an *automatic gain control* (AGC) system for this, but professional cameras usually provide a series of fixed gain settings that can be chosen by the operator. Since the circuits following the input amplifiers have a fixed dynamic range determined by the bits per pixel value, increasing the amplifier gain inherently means that the input light level must be reduced to avoid overloading the circuits. Increased gain also means that the SNR is degraded. This makes a trade-off between camera sensitivity and SNR.

Cameras with AGC (described in Section 5.3) perform the sensitivity-SNR trade-off according to the input light level to the camera. When the light level causes the lens iris to open fully, the amplifier gain will automatically be increased to maintain normal signal levels at the output. This is simple for the camera user but it does not necessarily provide the greatest artistic flexibility. Providing manual gain controls allows the professional user to choose iris setting, gain, and light level to accomplish the scene objectives. In professional cameras, using higher-than-normal video gain is called *high-sensitivity operation*. It is also common to reduce aperture correction or other enhancements in high-sensitivity mode to improve SNR. In some cases, for extreme sensitivity, the video bandwidth may even be reduced by the insertion of low-pass filters in either the analog or the digital domains.

The amount of gain above normal for high-sensitivity mode is a camera specification. It is given in dB: +12 dB gain means that the video gain is 12 dB (4×) above the normal level and the camera can produce normal output levels at 4:1 lower light levels. Most cameras have at least +18 dB of extra gain; some have as much as 30 dB. In the latter case, bandwidth reduction is essential to control the noise level.

4.3.3 White Balancing

All cameras provide means for automatic adjustment of the relative gains in the color channels to maintain a specified chromaticity as system white ($R = G = B$). This is called *white balancing*. Consumer cameras do this very simply, while professional cameras provide varying degrees of operator discretion for when to do white balancing, what part of the scene will be viewed as reference white, and so forth. It is important that white balancing be done early in the signal path because many other operations depend on the signals being matched.

White balancing can be accomplished by adjusting the analog input amplifier gains or it may be done digitally at a later point in the system. The latter approach requires that a certain part of the digital dynamic range overhead be devoted to white balance adjustments. Since white balance depends on the illumination color temperature, it is desirable to perform partial balancing with optical color filters to maintain the operation of the CCDs reasonably matched. Professional cameras have filter wheels for this purpose, but consumer cameras may not. In the latter case, all white balancing is accomplished by gain control, which is the lowest cost approach. However, that can lead to trade-offs in SNR and dynamic range, which is why professional cameras use more elaborate techniques. Automatic white balancing systems are discussed in Section 5.1.

4.4 HIGHLIGHT HANDLING

Highlights are areas of the scene where the light level is above the level chosen for scene reference white. These may come from illuminated objects such as light bulbs in the scene, or they may be highly reflective objects (mirrors—called *specular* highlights) that directly reflect a light source into the camera lens. The most demanding scenes for highlight handling occur outdoors in bright sunlight or at night when contending with bright artificial light sources such as car headlights.

Under certain circumstances, highlights may go many times above white level, as much as 100×. CCDs are designed to gracefully handle highlights that go above full-well conditions, but that does not mean there are no artifacts. Extreme highlights on a CCD may smear, bloom (grow larger), or show unusual motion artifacts.

The strategy of providing extra dynamic range in the CCD and the camera circuits for handling of highlights was introduced in Section 4.2.1.2. By setting system reference white level substantially below full-well, a degree of overhead is provided where the signal electronics can determine how the highlights are processed. A common approach is to compress excess highlights by having a "knee" in the system transfer characteristic to reduce incremental gain above a certain level. That allows the highlights to be reduced to fit the system dynamic range, without losing all the information. Figure 4.4 shows such a highlight compression approach. With this approach, it is possible to get a good picture of an indoor scene showing shadow details while retaining some detail in a window looking out into a brightly lighted outdoor scene.

By positioning the knee and setting the slope above it, the degree of visibility of detail in the highlights is controlled. The figure shows the case where the CCD saturation (peak white level) is set at 600% of reference white video level; this is determined by the video amplifier gains. The highlight compression knee is a little below 100% video level, so there is a range where highlights are compressed but some detail will still show. In addition, a white clipping level can be set for full suppression of further highlights—this is at about 110% of video level in the figure. In professional cameras, the knee position, knee slope, and white clip levels are usually adjustable.

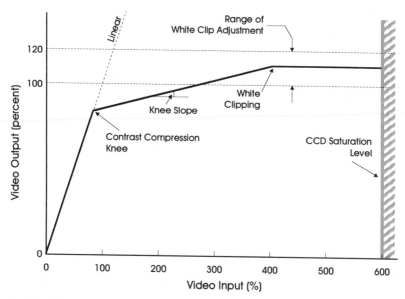

Figure 4.4 *Highlight compression.*

Some cameras have automatic knee operation that maintains the peak white in the scene at 100% while adjusting the knee to maintain the 100% light levels of the scene always at a specific point, say 90% output level. That provides the best possible reproduction of the scene brightnesses up to 100% level while allowing the last 10% of output dynamic range for display of highlight details. The output signal never exceeds 100%, which is important to digital video recorders that have a fixed limit of dynamic range in their ADC.

It is important to maintain color balance of the channels throughout the transfer characteristic so that operation of highlight compression does not affect colors. However, color saturation is reduced in the knee slope region by the same factor as the highlight compression. In some cameras, enhancement of saturation is provided on highlights to reduce the loss of color in compressed highlights. Of course, once white clip levels are reached, all color will be lost and the highlight will appear white. (If the highlight was strongly colored to begin with, white clipping will be reached first in one color channel.) But as the highlight level is further increased, clipping will occur in the other channels. Thus, the highlight will move closer to a full white. This effect cannot be compensated except possibly to force all highlights that go above white clip level in any channel to become white.

Another highlight compression option is to control image enhancement separately during knee compression. That allows the amount of fine detail in highlights to be modified.

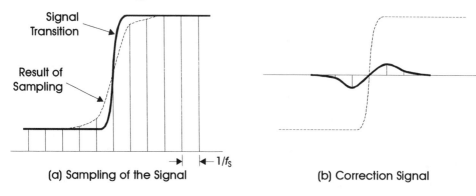

| (a) Sampling of the Signal | (b) Correction Signal |

Figure 4.5   *Sampled MTF loss and correction.*

4.5 IMAGE ENHANCEMENT

MTF loss caused by the sampling width of CCD pixels may be corrected both horizontally and vertically by *image enhancement* techniques. The basic idea is shown in Figure 4.5. Figure 4.5(a) shows a black-to-white signal transition being sampled. The length of the vertical lines represent the sampled values. As sampling approaches the positive transition, sample values begin to be affected somewhat by the signal value after the transition, so they become too large. Similarly, after the transition has been sampled, a few subsequent samples are affected by the signal value before the transition, so they become too small. The result is a smearing of the transition, shown by the dotted line in Figure 4.5(a). This is the effect of the MTF loss caused by sampling pulse width described in Section 3.4.1.1.

Figure 4.5(b) shows the type of signal that must be added to the sample values to correct the loss. This type of signal can be produced for either horizontal or vertical correction by digital high-pass filters. Horizontal filtering requires storing a number of adjacent samples and appropriately combining weighted values to produce the output. Vertical filtering requires storing a number of scan lines (usually two) and combining weighted sample values from each stored line for the output. These are often called *contour* or *detail* signals.

If the only correction was for the MTF losses inherent in the CCD's pixel sizes, the image enhancement could be given fixed values. However, enhancement also may correct for optical problems or it may precorrect for losses downstream from the camera, especially in video displays. Thus, the degree of enhancement is almost always operator adjustable.

4.5.1 Deriving Enhancement Signals

To conveniently adjust enhancement without affecting other parameters such as video gain, the usual architecture separates the steps of creating enhancement signals and applying them to the signal path. Figure 4.6 is a block diagram for a typical approach in an

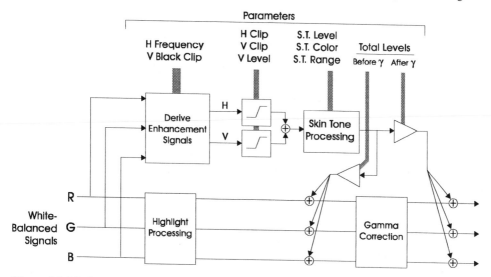

Figure 4.6 *Block diagram of a typical image enhancement system for a three-CCD camera.*

RGB camera.

Enhancement signals are derived from R, G, and B signals taken from the signal path after the channels are white balanced. Horizontal and vertical correction signals are created separately. In the case of horizontal correction, there may be a choice of the peaking frequency; a low frequency gives a coarse enhancement and a higher frequency gives a finer enhancement. Selection of the frequency depends on the nature of correction required for defects existing in the system beyond the camera. If only the camera MTF is being corrected, a high horizontal correction frequency would be chosen.

For a camera used in a composite video system such as NTSC or PAL, the horizontal detail correction should include a comb filter that prevents enhancement signal generation for video frequency components near the color subcarrier sideband frequencies. That prevents enhancement from causing an increase in luminance-chrominance crosstalk in composite systems.

As explained earlier, vertical enhancement signals are created by processing pixels that are adjacent in several scanning lines. Except for choice of the number of lines that will be processed, there is no variability in this. Usually the number of lines is fixed at the value three, which requires storage of two lines of information. (The current line is processed along with two previous lines.)

4.5.2 Processing of Enhancement Signals

Horizontal and vertical enhancement signals are processed both linearly and nonlinearly to accomplish the maximum enhancement while avoiding enhancement of noise in the picture.

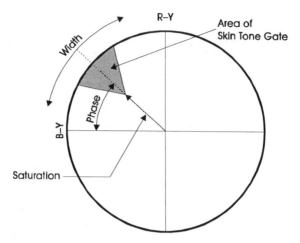

Figure 4.7 *Development of the skin tone gate signal (shaded area) from the color-difference signals. (Courtesy of Sony)*

Linear processing is simply the setting of the amount of horizontal and vertical correction to apply. Nonlinear processing usually includes a "dead zone" that prevents enhancement of noise or other small-amplitude variations. Enhancement only applies to signal changes that are above a certain threshold. The better the camera's SNR, the smaller the dead zone can be. Amplitude limiting or a knee function is also included in the correction signal path to reduce the correction of large amplitude transitions that could result in signal excursions exceeding the dynaimc range of later circuits. Finally, the correction signal path may be modified by the highlight-limiting circuits of the camera to control the degree of enhancement applied to highlights that have been compressed.

4.5.3 Skin Tone Enhancement

Some cameras have a special form of enhancement processing that allows a different degree of enhancement to be used for skin tone colors. Comparators on the color component signals determine when the color and luminance are in a range that may be human skin tones. This process is shown in Figure 4.7. A "skin tone gate" signal is created when the chrominance is within the shaded range shown in the figure. Some cameras have a feature for displaying this gate superimposed on the video signal so that adjustments can be made to keep the skin tone gating within the desired regions of the picture (only on the performer's faces.) This display is independent of the main video output from the camera so adjustments can be made while actially on-air with the camera.

Ordinarily, the skin tone gate information is used to locally reduce the application of enhancement signals or even to deenhance to prevent excessive visibility of facial details. It electronically does the same thing as applying makeup to the performer's faces.

By making this type of circuit adjustable to any range of colors (not just skin tones,) it may be used to enhance or deenhance any region of the picture. This can be useful for signals that are to be compressed; for example, a region of fine detail that is unimportant to the overall scene, such as grass, could be deenhanced to reduce the number of bits the compression algorithm would allow to grass (see Section 4.10).

4.5.4 Applying Enhancement Signals

There are several choices for how the enhancement signals are introduced back into the signal path. In an RGB system, the enhancement signals must be applied equally to each channel so that chrominance is not changed. On the other hand, in a YC_RC_B system, the enhancement signals may be applied only to the luminance channel.

It also makes a difference *where* the enhancement signals are added to the signal path. Introducing enhancement before gamma correction applies the same enhancement at all gray scale levels, but the enhancement of signals near black will be magnified by the action of gamma correction. Introducing the enhancement after gamma correction tends to favor enhancement of details in the white region of the gray scale. Many cameras provide separate adjustment of enhancement insertion both before and after gamma correction.

4.6 COLOR CORRECTION

The color response of a camera—its colorimetry—is determined by the camera manufacturer and generally is not adjusted in the field. As explained in Section 2.2.1, cameras use precise intermixing of the color channels (masking) to produce the negative lobes of the spectral response taking curves. This is a fixed linear matrix processing the color channels according to equations of the form:

$$R' = C_1R + C_2G + C_3B \tag{4.1}$$
$$G' = C_4R + C_5G + C_6B \tag{4.2}$$
$$B' = C_7R + C_8G + C_9B \tag{4.3}$$

Where the constants C_1 to C_9 can have positive or negative values.

Some cameras have adjustable masking for precise matching of colorimetry between several cameras. Such adjustment is difficult and requires great skill and special test images. It is seldom used except in telecine cameras where it is needed to correct for film comorimetry problems (see Section 7.7.3).

4.7 SHADING CORRECTION

As explained in Section 3.4.9, CCDs may have variation of sensitivity or black level over the area of the image. These effects come under the heading of *shading*. There may be gradual changes across the image or there may be abrupt variations from pixel to pixel. CCD performance in this regard has been steadily improving, but some cameras include electronic correction circuits for shading.

4.7.1 Black Shading

Variation of CCD dark current from pixel to pixel causes black shading. In high quality CCDs, this effect is in the range of 1% to 2%, although it is a strong function of temperature and increases rapidly as temperature increases. Since black shading is an additive component to the signal, it may be corrected by subtracting a corresponding correction signal.

Black shading is easily observed by capping the camera and observing video signals with a waveform monitor or picture display. In the digital domain, the same process can be used to capture a correction signal into a memory. Some cameras have a full frame memory that can store a shading correction value for every pixel. This memory is loaded by capping the camera and storing the video signal into the shading memory. In operation, the shading memory is read out in synchronism with the CCD scanning and the stored value for each pixel is subtracted from the video path. This is an elegant approach and, with the continued reduction of memory costs, it is not very expensive.

4.7.2 White Shading

White shading is a variation of sensitivity over the surface of the imager. It is observed by focusing the camera on a uniformly-illuminated white screen. Shading may occur in the CCD itself, or it may be an optical effect in the light-splitter or lens. The most common optical effect is falling-off of sensitivity at the edges of the image, which is called *vignetting*.

White shading may be corrected by modulating the video gain in accordance with the inverse of the curve of sensitivity across the image. Correction signals traditionally were simply sawtooth and parabolic waveforms at horizontal and vertical scan rates. These signals were adjusted manually or they were generated by a computer under a specified setup condition. Because the modulation required by shading correction is another factor that uses up digital dynamic range overhead, white shading correction is usually done in the analog domain before the ADC.

Today, with 12-bit digital processing that provides sufficient digital dynamic range, a three-dimensional zonal correction can be implemented in the digital domain, which allows more complex shading topographies to be corrected.

4.8 GAMMA CORRECTION

The gray scale response of a CCD imager is quite linear over a wide range of input, which means the video output is a linear function of the scene light level, right up to full-well conditions. However, the light output versus input voltage characteristic of a cathode ray tube (CRT) video display is nonlinear. Television standards, such as NTSC and PAL, call for nonlinear compensation in cameras to precorrect for CRT nonlinearity. This still exists today even though other display devices having different characteristics, such as LCDs

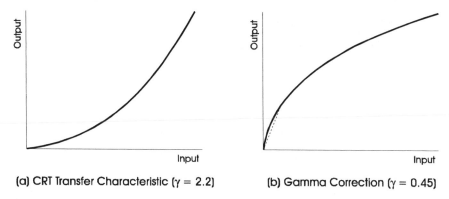

(a) CRT Transfer Characteristic (γ = 2.2)   (b) Gamma Correction (γ = 0.45)

Figure 4.8 CRT gamma and its correction.

or plasma panels, are coming into use. The process of correcting gray scale response is called *gamma (γ) correction*.

4.8.1 Specifying Gamma

Gamma is usually specified in terms of an exponent applied to the amplitude transfer characteristic. For example, a CRT has a gamma of approximately 2.2, which comes from the inherent nonlinear control of an electron gun. Thus, a CRT transfer characteristic may be expressed as

$$B_O = V_I^{2.2} \quad \text{or} \quad B_O = V_I^{\gamma} \tag{4.4}$$

where B_O is the screen brightness and V_I is the input voltage. This also means that the CRT becomes more sensitive to its input signal as the brightness level increases. If that is not corrected, CRT displays will show little detail in dark regions of the picture. To compensate for the CRT gamma, the signal must be processed by an exponent that is the reciprocal of the CRT exponent: $\gamma = 1/2.2$ or 0.45. Figure 4.8 shows this relationship.

A more flexible approach to the gamma problem would be to require video signals to be linear and include appropriate gamma correction along with the display device that causes nonlinearity. That choice was not made in the early days of television because an objective was to minimize the cost and complexity of receivers by doing as much of the processing as possible at the transmitting end. That is not necessarily a good argument today; requiring gamma correction in receivers or monitors would increase their cost by a trivial amount.

There is another advantage of gamma correction at the transmitting end for systems that use amplitude-modulated transmission (as does terrestrial broadcasting of television) because the CRT gray scale characteristic reduces the visibility of transmission noise in the black areas of the picture. However, this is not a consideration for frequency-modulated or digital transmission.

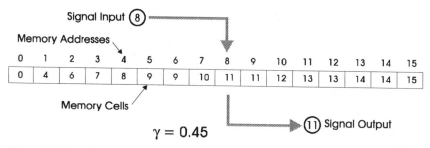

Figure 4.9 *Four-bit example of a memory look-up table for gamma.*

4.8.2 Gamma Circuits

Analog gamma circuits involve synthesizing the nonlinear curve with straight-line segments produced by diode circuits. These are difficult to set up and are prone to drift. Since gamma is required separately in each of the color channels and they must be closely matched to maintain color balance at all signal levels, accuracy and stability of gamma circuits is important.

Digital gamma correction is done with the look-up table approach, described in Section 4.2.2. A simple example of this is shown in Figure 4.9, which is a $\gamma = 0.45$ circuit for a 4-bit signal system. A 4-bit signal has an integer range from 0 to 15. The input signal is the address into the memory table; the output is the memory contents at that address. In the figure, an input value of 8 yields an output of 11.

If one examines the memory table contents, it will be seen that some amplitude resolution present in the input at higher levels is lost in the output because several adjacent memory cells have the same value. This happens because the reduced slope of the gamma curve at the higher levels requires more than 4 bits to be fully reproduced. It is an example of a place where the signal processing requires more bits per sample at the output to prevent loss of information. Notice that the opposite thing is happening on the lower range of the curve—with only 4 bits at the input, there is no output possible for the values 1, 2, and 3. That means the input also should have extra bits per sample to account for the gamma curve's expansion of the signal at low levels.

In Figure 4.8(b), it can be seen that the signal expansion (increase of slope) rises as the input level falls. Actually, for a true exponential function, the slope approaches infinity at zero input level. This is undesirable because the increase of gain causes enhancement of noise at low signal levels. Practical gamma circuits generally place a limit on the maximum gain at low levels; below a certain point, the curve is made into a straight line to prevent further gain increase. Figure 4.8(b) shows a dotted line that corresponds to a gain limit of 4:1. That is typical of the design of gamma circuits for 0.45 gamma.

Returning to the matter of bits per sample, a gamma-circuit gain limit of 4 means that the input dynamic range should be 4× more than the output range to prevent loss of output detail at low levels. That calls for 2 bits extra for the input bits/sample. Thus, a camera that will deliver 8 bps at its output, should have at least 10 bits/sample at the input to its gamma circuit.

4.9 SCAN CONVERSION

With digital processing, it is possible for the scanning format in the camera's imagers to be different from the scan format delivered at the camera's output. At first glance, this may seem wasteful, but there are several situations where it is advantageous to do scan conversion in the camera:

- One camera can support more than one type of video system.
- The picture quality of an interlaced system may be improved by the use of progressive scan in the camera. The conversion from progressive to interlaced can reduce vertical aliasing.
- An HDTV digital camera can deliver downconverted standard-definition TV (SDTV) digital component output with higher quality than is possible with SDTV cameras. Since HDTV and SDTV will coexist in the digital TV era, camera designers are providing both outputs from the same cameras.

Full scan conversion for both horizontal and vertical scanning is still an expensive proposition, but certain limited cases in the list above are quite practical. The best example is progressive-to-interlaced conversion (see Section 13.2).

4.10 VIDEO COMPRESSION

Digital video, whether component or composite, yields data rates upwards of 100 Mb/s. HDTV component systems require more than 1,000 Mb/s. Such rates are high compared to the capabilities of economical recording or transmission systems. Although digital recording systems that can handle the full rates are available for professional use, they are too expensive for many markets. Digital transmission of 100+ Mb/s is also available but requires significantly more bandwidth than analog transmission of the same quality of video. These problems are solved by the use of video data compression, which is one of the advantages of digital video systems, as listed in Section 1.5.1. This section gives an overview of video compression and its use.

Data compression [3] is widely used in digital computer systems for many kinds of data. In most cases, the data must not be changed by the compression-decompression process, a requirement called *lossless* compression. That limits the degree of compression possible. With video data, however, precise data reproduction is not strictly necessary—if the reproduction is good enough that viewers cannot *see* any artifacts in the video, it is satisfactory. This is *lossy* compression, which allows much higher compression factors.

Determining the effectiveness of video compression only by what a viewer can see is a dangerous criterion. Some viewers will see more than others. Preferably, evaluation of lossy compression should be reduced to hard numbers by suitable measurement techniques. The industry has not yet achieved this. Even more importantly, some video processes that may occur after lossy compression has been done may be more sensitive to artifacts than a human viewer is. This is the reason that many video professionals do not want to use compression at all.

4.10.1 Compression Basics

The scanning process inherently generates a lot of redundant information in video signals. Adjacent pixels, both horizontally and vertically, repeat the same information in picture areas of solid color. In other areas of the picture, adjacent pixels may not be repeating, but they may be related in simple ways. The objective of *spatial* video compression is to find these relationshipe between pixels and transmit them more efficiently.

Scanned video transmits 30 or more frames per second. There is repeated information in areas of the picture where nothing has moved from frame to frame—this is *temporal* redundancy. The objective of temporal compression is to find areas of a frame that have not changed from the previous frame and simply send some codes that tell the display to retain the previous information. An extension of this technique is to look for areas that have moved but the data exists at a different location in a previous frame. In this case, the system can send a motion vector that tells the receiver where to find the information in the previous frame.

With digital video techniques, both spatial and temporal compression are practical. However, the sophistication is limited by the amount of processing power available at each end of the system—encoding and decoding. In many systems, it is reasonable to devote massive processing power at the originating end (encoding), because that happens only once to a particular signal and an expensive processor may be acceptable. It may even be practical to take extra time to perform compression, if it will allow a better job to be done.

At the receiving (decoding) end of the system, there may be many receivers, which have to decode in real time, and the cost of decoding must be strictly limited. Therefore, the decoding task must be simplified. A system that has elaborate encoding and simple decoding is called an *asymmetrical* compression system.

The asymmetrical concept is no help to a camera because the camera is always at the encoding end. A camera that delivers compressed video must contain the full encoding processor and it must operate in real time. Because of this and other needs, integrated circuits have been developed to perform compression in real time and at reasonable expense.

4.10.2 Compression Standards

Digital processing is extremely flexible and offers essentially an infinite variety of compression possibilities. However, no single receiver could decode all the possibilities, so there is a need for standards. Fortunately, this can be do so that the digital flexibility still allows a considerable degree of variability within a standard. Compression standards typically offer options to suit different applications.

A complete description for a set of compatible encoding and decoding techniques is called an *algorithm*. A standard may include several algorithms as options. Although there are many compression standards available, especially in the personal computer (PC) field, the two most important ones are worldwide standards developed by a collaboration of the *International Standardizing Organization* (ISO) and the *International*

Electrotechnical Commission (IEC). One of these, intended for still-image compression, was developed by an ISO/IEC subcommittee called the *Joint Photographic Experts Group* (JPEG) and is known as the JPEG standard (pronounced "J-peg.") The other, for motion video compression, was developed by the *Moving Picture Experts Group* (MPEG) of the ISO/IEC, and is called the MPEG standard (pronounced "M-peg.") The MPEG and JPEG standards are being applied in HDTV standards, the new DVD optical disc standard, satellite broadcasting to the home, digital still-picture cameras, video recorders, and video cameras.

It must be made clear that video compression—as good as it is—is not always suitable, especially in video production and postproduction situations. This is because compression subverts the idea that a digital video system should be transparent. Compression is based on the idea that redundancy may be removed from the signal without affecting the final subjective picture quality. If the viewer cannot see it, a compression algorithm can safely remove it. However, something that a viewer cannot see may be important to a postproduction processor and the output of the processor will be affected in a way that viewers will see.

More importantly, many postproduction effects processes can only be done with signals that are in component form—uncompressed. For a compressed video system, this means that decompression must be done before effects processing and the signal must be recompressed after the processor. This can lead to repeated passes through the decompression-recompression process, which may result in accumulated distortions. In a sense, compression is reintroducing some undesirable analog properties into digital systems. For example, high-end video systems tend to have no video compression at all, which makes them expensive. To compensate, products are now being introduced with minimal compression algorithms designed to prevent accumulated distortions, but still allow the user to enjoy some of the cost savings of compression.

4.10.3 Spatial Compression—JPEG

Redundancy in still images may be identified by comparing pixels and lines, looking for similarities or patterns. This process is simplified by handling pixels in blocks, for example, 8×8 or 16×16 pixels at a time. When redundant patterns are found, they may be coded by a more simple means for transmission. The two steps can be combined with the use of a *block transform* that takes a block of pixels and converts it to a different form that is more easily compressed. The most popular such transform is the *discrete cosine transform* (DCT), which is used in the JPEG [4] algorithms. The DCT converts a block of pixels to an equal-sized block of coefficients that represent the horizontal and vertical frequency components of the block. This is shown in Figure 4.10.

An 8×8 block of pixel values is transformed to an 8×8 block of frequency coefficients as shown in the figure. The coefficients represent the amplitudes of increasing spatial frequencies in the block. This is amenable to two types of compression:

1. Recognizing that higher spatial frequencies are less visible to a viewer, the quantization of the higher-frequency components can be made coarser, thus saving

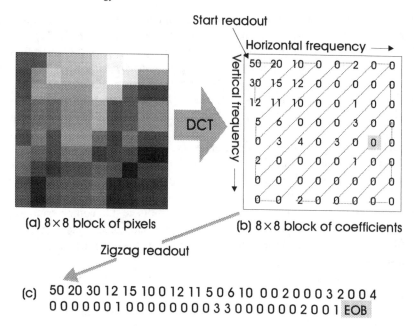

(a) 8×8 block of pixels

(b) 8×8 block of coefficients

(c) 50 20 30 12 15 10 0 12 11 5 0 6 10 0 0 2 0 0 0 3 2 0 0 4
0 0 0 0 0 0 1 0 0 0 0 0 0 0 0 3 3 0 0 0 0 0 0 2 0 0 1 EOB

Figure 4.10 *The discrete cosine transform. (From [8]).*

    bits. Coarse quantization also makes many of the small-amplitude coefficients go to zero.

2. If the coefficients are read out in a zigzag order shown in the figure, which produces a sequence of increasing frequency, most of the information content appears early in the sequence and, after a point, all further coefficients are zero. This is ideal for *run-length coding*, which replaces repeated values with a single value and a count number. The *end-of-block* code (EOB) signifies that all coefficients beyond it are zero.

The performance of DCT compression is controlled by specifying a quantization table that tells how many bits to use for quantizing each frequency component. This makes a trade-off between picture quality and degree of compression. With good picture quality for most applications, JPEG, using DCT, can compress image data by 10:1 or more. Higher compression factors are possible in situations where lower image quality is acceptable.

4.10.3.1 JPEG Decompression

Use of a JPEG-compressed image requires decoding of the run-length and other coding to obtain the quantized frequency components and then, the application of an *inverse DCT* process to recover the original pixel blocks. Because of the quantization, this is inherently a lossy process. The DCT process and its inverse are very compute-intensive. in fact, they are about the same, meaning that JPEG is a symmetrical algorithm. Integrated circuits are available for JPEG compression or decompression, so the standard can be widely applied without resorting to the details in each application. In PCs, it also can be implemented in software; with today's fastest PCs, JPEG compression or decompression of a high resolution image takes only a few seconds.

4.10.3.1 JPEG Options

The JPEG standard includes a wide range of algorithms, both lossless and lossy, for applications ranging from document storage and retrieval to high-quality photographic image reproduction to electronic still-image cameras. The lossless algorithms cannot use DCT because it is inherently lossy. The lossy options are of most interest for video.

The are four modes of JPEG operation:

- Lossless—the image is reproduced exactly. All other modes are lossy.
- Sequential—the image is encoded in the order it was scanned. This is the normal mode.
- Progressive—this mode first transmits a coarse image that can be rapidly displayed at the receiver. It is followed by repeated encodings at progressively higher resolutions. This is useful in a situation where the user wants to quickly see an image without waiting for the transmission of a full high-resolution version of it.
- Hierarchical—this mode encodes the image at multiple resolutions; the user may choose which one to view.

4.10.4 Motion Video Compression—MPEG

In principle, motion video can be compressed by performing spatial compression on each frame, using an algorithm such as JPEG. This is actually done and an algorithm known informally as *motion-JPEG* (M-JPEG) is often used. It has the advantage of widely available ICs and, most importantly, all frames remain independent of all other frames. Of course, more advanced motion compression schemes exploit the redundancy between frames, making frames not independent of one another. In video editing, for example, that prevents editing at arbitrary frame locations. M-JPEG is often used in video editing systems, but it is not a standard and has been applied differently in different products. A better approach is to use a version of MPEG that does not use interframe compression (see Section 13.4.1).

4.10.4.1 Motion Compensation

The real advantage of motion video compression occurs in the application of temporal compression. The most widely-used technique for that is *motion compensation*, which is used in the MPEG standard [5]. Like DCT, motion compensation divides the image into x-y blocks of pixels. It also depends on storage of previous frames at the receiver. During compression, each block is compared to the corresponding block in the previous frame. If

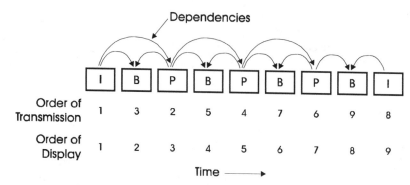

Figure 4.11 *Typical frame sequence for MPEG compression.*

there is a match, the new block is not transmitted, but only a code that tells the receiver to use that block from the previous frame.

If there is no exact match of blocks between frames, the new block is then compared to a range of nearby locations in the previous image to look for the possibility that the block may just be moved a little from the previous frame. If a slightly moved match is found, a *motion vector* code will be sent telling the receiver where to find the block in the previous frame. Finally, the blocks that were not matched are DCT compressed and sent by the same methods as in JPEG. Motion compensation typically achieves 3:1 more compression than spatial-only techniques. However, it is a substantial increase in the complexity of encoding.

4.10.4.2 MPEG Frames

Individual frames in an MPEG sequence may be compressed in one of three different ways

1. A frame that is compressed as a still image is called *intracoded* (I-frames). It does not depend on any other frame for its content. Video editing can always be done at an I-frame.

2. Frames that are based on differences from the preceding frame are called *predicted* (P-frames).

3. Frames can also be based on the previous and the next frame; these are *bidirectionally predicted* (B-frames). This form provides the most compression, but it requires more memory at the receiver and obviously, the "next" frame must be available at the receiver before a B-frame can be decoded. This constrains the sequence of transmission of frames. Not all MPEG coders use B-frames.

A typical sequence of frame types is shown in Figure 4.11. I-frames always must be transmitted at the start of a video sequence and one should generally be sent whenever there is a scene change in the video. In addition, it is good practice to send I-frames periodically to support editing or for recovery in case of an uncorrected transmission error. The latter is because a transmission error in one frame can be propagated to subsequent frames by the predictional decoding. On the other hand, too many I-frames increases the data rate.

Figure 4.11 shows how frames depend on each other and how the use of B-frames requires the transmission order to be rearranged so the base frames are always available at the receiver before a predicted frame is to be decoded.

4.10.4.2 MPEG Options

Since MPEG compression is used for continuous display of video, data *rate* becomes an important specification. Data rate is controlled by adjusting the compression parameters the same way that compression factor was traded for picture quality in JPEG compression. However, with fixed compression parameters, the data requirement will change from frame to frame as the picture detail and the amount of motion changes. Often, it is desired to generate a fixed data rate for transmission or recording and the technology of rate control is very important to MPEG systems.

Considering that compression becomes more difficult as the data rate target is reduced, several MPEG varieties have been developed for different points on the picture quality versus data rate curve. MPEG-1 was the first standard completed; it is designed for dats rates around 1.5 Mb/s and delivers picture quality comparable to a VHS home VCR. Its first use was for storing of video on CD-ROMs in PCs.

The MPEG-2 standard [6] came a little later and, for 525-and 625-line video, it is designed for data rates above 5 Mb/s. A high-quality version of MPEG-2 is used in the ATSC standards for HDTV in the United States. There, the data rate for HDTV is 19.3 Mb/s (see Section 13.4).

The MPEG-2 standard also includes a very flexible data transmission format that supports multiple audio, video, and supplementary data channels in a single data stream. This is a packetized format that is finding use even in systems that do not use MPEG video compression (see Section 13.4.2).

4.10.5 DIGITAL TV STANDARDS

Most countries in the world are planning a transition to digital broadcasting of television. Standards have been set in the United States after nearly 10 years of development; this standard is the *Advanced Television Systems Committee* (ATSC) [7] standard. It was developed by the *Grand Alliance*, a group of companies that had made proposals for such a standard under the direction of the FCC's *Advisory Committee on Advanced Television Systems* (ACATS). The standard recognizes that digital systems can support a variety of scanning standards and it provides for both SDTV and HDTV transmission over the same broadcast channel. The compression is MPEG-2, but a special transmission link format is defined using MPEG-2 packetization for transmission of a flexible combination of SDTV channels, an HDTV channel, and multiple audio channels in the same 6-MHz channel bandwidth. This subject is covered further in Chapter 13.

In Europe, a slightly different digital broadcasting standard has been developed, called *Digital Video Broadcasting* (DVB). This standard includes modes for terrestrial broadcasting, satellite broadcasting, cable, and recorded-media distribution. It also uses MPEG-2 video compression but the audio format is different and a different transmission data stream format is used. DVB is proposed as a worldwide standard and it is already in use for direct-to-home satellite broadcasting around the world.

4.11   SIGNAL ENCODING

There are four possible types of encoding for the output of a video camera:

1. A serial bitstream;
2. Components, analog or digital;
3. Analog composite—NTSC or PAL;
4. To a recorder.

Details of these formats are covered in other chapters; only the signal processing considerations are covered here. As in the rest of this chapter, the camera is assumed to have digital processing.

4.11.1   Serial Bit Stream Encoding

Serial encoding is a viable approach when a camera must deliver output to a digital video system. Serial encoding requires consideration of data synchronization, identification, and clocking. These issues are handled by the serial format standards, covered in Section 6.4. The subject here is the matter of getting the video data from the multichannel format of the camera processing to a single serial channel that can be plugged into the standard format.

Camera processing in general is organized into a parallel format with separate circuits for each bit of the samples passing through the system. Thus, for an RGB component camera operating at 10 bps, there are 30 parallel channels of data coming from the camera's processor. If the camera processing operates at 18 MHz, which is typical for a 525- or 625-line camera, a serial data rate of 540 Mb/s is required. Most cameras would not output this data rate but would instead convert the RGB format to YC_RC_B, which might then be resampled to 8-bit samples at 13.5 MHz. This creates a 4:2:2 component signal that reduces the raw data rate to 216 Mb/s without any significant loss of picture quality. That rate can be handled by the Society of Motion Picture and Television Engineers (SMPTE) 269M serial interface standard (see Section 6.4.2), which uses coaxial cable operating at a total rate of 270 Mb/s. The additional data rate provides overhead for synchronization, identification, and error protection.

The actual steps of getting from 10-bit parallel RGB to the serial interface is shown in Figure 4.12. The interleaving of samples is according to the sequence $Y-C_R-Y-C_B-Y$. . . as specified in the SMPTE 125M parallel interface standard. In serializing, all the bits of each sample are transmitted sequentially, with the least significant bits (LSBs) coming

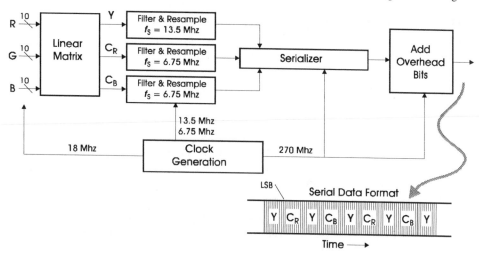

Figure 4.12 *Conversion from 10-bit parallel RGB to SMPTE 259M serial transmission.*

first, as shown in the figure.

The resampling process to reduce the sampling rate is an interpolation of samples that is accomplished with a digital filter. Filtering is necessary to reduce the bandwidth of the incoming samples to remove components above the Nyquist limit of the new (lower) sampling rate. This process of rate reduction may seem to be a waste of information, but the 13.5 MHz rate has sufficient bandwidth for 525- or 625-line systems. The use of excess sampling at the input provides reduction of aliasing.

4.11.2 Component Parallel Encoding

The output from the camera's signal processing is 10-bit RGB components and outputting that format would seem to be the simplest thing to do. However, that amounts to three 10-bit channels, requiring a 30-bit parallel interface, which is quite impractical. SMPTE Standard 125M provides a 10-bit parallel interface by interleaving the three component channels for each bit in the sequence described in Section 4.11.1. This interface reduces to 10 parallel channels, which can use a cable with 12 twisted-pair conductors. That is fairly expensive and awkward, but it is workable for short distances. There has not been much use of this interface; most applications have gone to the serial interfaces, which have much less expensive cabling and can cover longer distances.

4.11.3 Analog Composite Encoding

Very few video systems are yet all-digital and most digital cameras are called upon to provide analog composite outputs. Encoder chips have been designed to go directly from digital component signals to a digitized composite format in either NTSC or PAL. Passing the output of the chip through a DAC yields a completely standard analog composite signal. Because of the precision of digital circuits, such encoders actually give better performance, reliability, and stability than analog units.

4.11.4 Connecting a Camera Directly to a Recorder

The most common case of connecting cameras directly to recorders occurs inside of a camcorder. In that case, the interface can be anything, even 30-bit parallel, because distances are short and cabling is not an issue. However, the recorder itself will require some serialization (see Chapter 8), and it is reasonable to begin that process in the camera.

Professional camcorders are available with digital component recording, but lesser camcorders generally record compressed video to save size, weight, and cost. Depending on the application of the camcorder, more or less compression may be used. For example, the *Digital Video Cassette* (DVC) system, which was originally designed for home and semi-professional use, compresses 525- or 625-line video to about 25 Mb/s. A professional version of DVC, called DVCPro, compresses only to 50 Mb/s to provide better quality and applicability to editing.

REFERENCES

1. Luther, A. C., *Principles of Digital Audio and Video*, Artech House, Norwood, MA, 1997.

2. Ibid., Section 5.2.2.6.

3. Ibid., Chapter 9.

4. ISO/IEC Standard IS 10918-1, *Digital Compression and Coding of Continuous-Tone Still Images.*

5. ISO/IEC Standard IS 11172-3, *Coding of Moving Pictures and Associated Audio for Digital Storage Media at up to 1.5 Mbits/s.*

6. ISO/IEC Standard IS 13818-1, -2, *MPEG-2 Systems, MPEG-2 Video.*

7. Advanced Television Systems Committee (ATSC), http://www.atsc.org.

8. Luther, A. C., *Principles of Digital Audio and Video*, Artech House, Norwood, MA, 1997, Section 8.3.2.

5

Automatic Camera Operation

In addition to the signal processing that recovers signals from the CCDs, optimizes them, and formats them for output, cameras contain numerous other electronics features that provide automatic operation and convenience. In this book, these are camera automatics, and they include

- Automatic white balancing
- Automatic focusing
- Automatic exposure
- Image stabilization
- Electronic zoom

and possibly others. Of course, there are also important system features in professional cameras; these are covered in Chapter 6, and mechanical features are covered in Chapter 8. This chapter discusses features that are predominately electronic.

5.1 THE SIGNIFICANCE OF AUTOMATIC OPERATION

Many of the features discussed here automate basic camera operations, such as focus, exposure, and white balancing. Early television cameras did not have such features and it was up to a skilled cameraperson to keep them under control. In professional video production, these parameters are often considered to be artistic aspects of camera operation and automating them is undesirable. However, in the consumer market, these same features make possible successful camera operation by unskilled consumers; without such features, that market could not exist.

Automatic features have been highly developed for consumer cameras, and they have become capable enough that they are now considered desirable for certain professional operations such as news gathering, where the camera operator has enough to do getting to the scene and keeping the camera aimed properly, without worrying about focus, exposure, and color balance. In professional studio operation, however, automatic features are seldom used.

5.2 AUTOMATIC WHITE BALANCING

The objective of white balancing is to make a color that appears "white" in the scene appear "white" when reproduced on a video display. That does not mean the two "whites" are identical; it simply means that the viewer will perceive each to be white under the current local viewing conditions.

"White" is easily defined in systems that use color difference signals, such as Y, C_R, C_B components, or NTSC or PAL composite. It is the condition where both color difference signals are zero or, for composite, the color subcarrier amplitude goes to zero.

The definition of white is more difficult for RGB component systems. Usually it is taken as the condition where R = G = B. However, that depends on the system being balanced so that equal RGB signals do result in white at the display. This condition must be deliberately set up and maintained at any point in an RGB system where color balance is important. That is easily accomplished in a digital system by establishing white balance at the input to the ADCs. Then, white will correspond to equal digital values in all channels. Since each digital process that follows can be precisely defined for all channels, color balance will be maintained through the system.

5.2.1 Detecting White Balance

There are two questions for white balancing: (1) what part of the scene is white, and (2) when to adjust the balance. The first parameter is easily handled if a white card is placed in the scene and viewed by the camera for white balancing. That, of course, is only practical if the scene is accessible and can be disturbed by placing the white card. Since the actual chromaticity of "white" depends on the color of the scene illumination, balancing cannot be always be done by turning the camera away from the scene for balancing on a test chart. That works only if one can be assured the test chart illumination color is the same as the scene's.

Since scenes do not always contain white objects and such objects may move around, accurate white balancing cannot be a continuous process. It must be activated under operator control when it is known that a white object is in a suitable place in the camera's field of view. This is usually managed by placing a cursor on the camera viewfinder screen that locates the area where the camera will sense for white during the balance procedure. The operator manipulates the camera and/or the scene to place a white object within the cursor's area and then activates the balancing procedure. This is not fully automatic in that the camera operator must interrupt shooting and perform the procedure every time a white balance adjustment is needed. However, it is the only reliable way to do it.

Consumer cameras, however, do often use continuous white balancing. By selecting the peaks of video in each channel and comparing channels to determine where they all

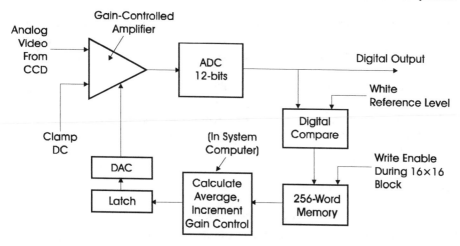

Figure 5.1 *White balancing in a digital camera.*

peak at the same time, a good approximation to white can be determined. White is assumed to be at the places where the three channels peak together. It is then a simple matter of balancing the signals at those points (see Section 5.2.3).

5.2.2 Deriving Correction Signals

A digital white balancing system that uses analog gain control is shown in Figure 5.1. Reference white occurs at a digital level of 900 on a 12-bit ADC. That allows more than a factor of four for highlight overhead. A comparator derives the difference between the ADC output and the white reference level, 900; that difference signal is gated into a memory during a 16×16 pixel (256 total samples) white-sampling block. The memory is capable of accepting the samples as fast as they are generated by the ADC, wheras the computer cannot process samples that rapidly. Sometime before the VBI, the computer reads the 256 samples from the memory and computes an average. The averaging is necessary to reduce the effects of noise and shading on the balancing.

The average information is used to increment the gain value in the proper direction. The new gain setting is output by the computer to a latch circuit that holds the value for the analog amplifier gain control through a DAC.

During the balancing procedure, the computer adjusts the gain by one step in each VBI until the average of the comparator output goes to zero. With gain steps of 0.25%, balance can be achieved in about 5 seconds. In an RGB camera, this process is performed separately in each channel.

5.2.3 Continuous White Balancing

Some home cameras have a white balancing system that is fully automatic—no button has to be pressed to enable a special balancing mode. Instead, the camera tries to figure out what is white in the scene and continuously balances on that. Obviously, such a system cannot do the right thing if there is not any real white in the scene; in such cases the best thing it can do is to hold the previous balance condition until some white appears in the scene.

These systems examine the signal components for areas of high luminance and relatively low chrominance. Such areas are probably white and can be used for balancing. The balancing circuit continuously searches for white areas and, if enough are found, it proceeds to balance on them. It is an approximation, but it can be effective under most conditions.

5.3 AUTOMATIC EXPOSURE

The video output level is controlled by several factors. These are:

1. Exposure—this is usually associated with adjustment of the lens iris, which controls the light level on the CCDs.
2. Neutral-density filters—these may be placed in the optical path to extend the exposure range for brighter scenes. They are also useful for normal scenes where it is desired to open the iris up for reduction of depth of field. Professional cameras have a filter wheel in the optical path between the lens and the prism for ND filters.
3. Video gain—this is the amplifier gain between the CCD outputs and the ADCs. Increasing the gain reduces the CCD output needed to reach full video level, thus requiring less light on the CCDs for increased camera sensitivity. The trade-off is that video SNR is reduced with higher gain.
4. CCD shuttering—electronic shuttering of the CCDs, which is used to shorten exposure *time* for sharper reproduction of moving objects, also reduces camera sensitivity.

An automatic exposure control system must consider all these factors in its algorithm. Because there are various ways in which these factors may be combined for different objectives, a versatile exposure system may provide several modes of operation. The ultimate cop-out is, of course, manual control, and all cameras should provide that so the operator can take complete control when necessary.

5.3.1 Detecting Video Level

Digital sensing of video level is much the same as the detection of level for white balancing, described in Section 5.2.1. A comparator compares video peaks to a digital reference value. The result is used to control the system. However, that is too simple because an important question has to be answered: What are video peaks?

Literally, "peaks" of video level are the points of highest level. But the highest levels may well be highlights that should be compressed or clipped to produce the best reproduction of the desired areas of the scene. If the level control system brings such peaks

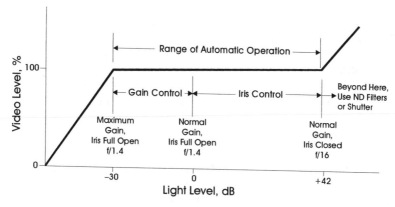

Figure 5.2 *Automatic exposure algorithm using both iris and gain control.*

down to normal levels, most of the scene will be crushed into the blacks. A detecting system must provide for a compromise that will allow highlights to go over 100%, while controlling only from the desired parts of the scene. This is an artistic judgment that is not always possible to be made by an automatic system. It can be quantified by expressing the amount of picture area contained in highlights versus the area contained in true white peaks.

One approach for an automatic system is to compute a histogram of the percentage of scene area at each video level. The user can then be given a control to specify the percentage of area to set to reference level. Higher levels will go over reference level and be handled by the highlight compression system. It could be argued that this is not really an automatic system because there is still a user control. However, that control is set for the general character of the scene, after which the automatic system takes care of variations in illumination or scene content. The amount of operator attention to exposure is much reduced and the result is much better than pure manual control.

Home cameras generally do not provide such a control; there is just a choice of on or off for the automatic exposure. Manual control may have to be selected to obtain good reproduction of scenes containing large areas of highlights.

5.3.2 Automatic Exposure Loops

Having detected peak level in the signal, a further question arises: What to adjust to control the level? There are four possibilities—iris, gain, filter wheels, and shuttering. To cover the full range of camera capability, the automatic system would have to adjust them all. Generally, that is not done; filter wheels and shutter adjustments are left to the operator and the automatic exposure control works only with iris and gain. In fact, the most professional cameras generally also leave the gain setting to the operator and only the iris is controlled automatically. Therefore, the operator must manually choose the other settings that will result in the iris operating in the range desired. Figure 5.2 shows an algorithm for automatic exposure that controls both iris and gain. Gain is kept at nominal until the iris reaches full open. Further loss of light level causes increasing of gain. As the figure shows, this system will hold the video level constant over a 72 dB range of light level. Below that range, the signal output falls because the light is lower than the camera's maximum sensitivity and above that range, the lens is all the way stopped down. To operate at higher light levels, ND filters or the electronic shutter must be used to reduce sensitivity, thus bringing operation into the range of the automatic exposure system.

5.4 AUTOMATIC FOCUSING

Focusing is an example of a maximizing process; focus is adjusted for maximal sharpness. Because one cannot detect a maximum until it has been passed through and the value begins to reduce on the other side of maximum, automating focus requires a special kind of feedback loop. In addition, the same kind of questions come up as with exposure control: how to detect focus and what part of the picture to do it on.

5.4.1 Detecting Focus

Depending on the scene content and the depth of field, the whole picture may be in focus at once or, only a small part of the picture may be focused. An electronic detection system must decide what part of the scene to keep in focus. This may not always agree with the camera operator's desire. For that reason, there should always be provision for the operator taking over manual control when the automatic system does not do what he or she wants. The usual approach is for the electronic focusing system to examine an area in the center of the scene, under the assumption that the interest of the scene will always be at the center of the frame and that should be kept in focus.

In general, the high-frequency energy present in the luminance video signal is proportional to the sharpness of the image. As focus is adjusted, the high-frequency content will go through a peak at the point of best focus. However, high frequencies relate only to the sharpness of vertically oriented objects; if an image contains only horizontally-oriented objects, it does not have much high-frequency energy. Focus could still be detected by examining the vertical-resolution energy, but that is more difficult and is seldom done.

In a digital system, high-frequency energy is indicated by large differences between adjacent or nearby samples. A digital high-pass filter can be made by processing groups of samples. using an algorithm that cancels out similarities between samples and enhances the differences. The output of such a filter is gated to select the samples from the desired area of the frame and the total energy in the samples is summed to create a control signal.

5.4.2 Rangefinder Focus Detection

Early photographic cameras used a *rangefinder* system to measure the distance to an object in the scene. That information was coupled to a calibrated focusing mechanism of the lens to control focus. The photographer looked into the rangefinder and adjusted

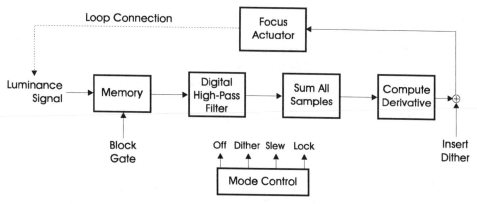

Figure 5.3 *Automatic focusing.*

focus of the lens until a split image of the scene converged on the object to be kept in focus. This was not automatic in that the photographer had to look at the rangefinder and manually close the focusing loop, but the same principles can be used for automatic focusing.

The idea of a rangefinder is to view the scene from two points separated by a certain distance d. The angle θ between the light rays from the two points viewing the same object can be used to calculate the object distance D according to

$$D = d / \tan(\theta/2) \tag{5.1}$$

Various methods have been developed to use this principle in video cameras. In general, the rangefinder output must be coupled somehow to the lens focusing mechanism. In the early photographic cameras, it was a mechanical link. In video cameras today, it is more likely to be electronic.

As described above, rangefinders have the problem of *parallax*—the lens views the scene from a slightly different angle than the rangefinder. This causes errors, especially at short object distances. Canon has developed a unique rangefinder system using infrared light that passes through the taking lens itself. That eliminates the parallax problem.

Another type of rangefinder uses an infrared radar approach. A beam is sent out toward the scene; it reflects from the nearest scene object and is detected by an infrared detector at the camera. As in radar, the time delay between the beam going out and its return is used to calculate distance.

5.4.3 Focus Control Loops

Having a signal that indicates the amount of high-frequency energy derived from the desired area of the frame, focusing can be accomplished with a maximizing servo operating on the time derivative of that signal. However, the control signal does not change unless the focus is changed, so there will be no derivative unless the focus is deliberately moved or dithered. But with some dither, the derivative is positive on one side of the maximum, goes through zero at the maximum point, and becomes negative on the other side of maximum. This is shown in Figure 5.3. The servo is designed to settle at the point where the derivative signal is closest to zero.

Various other features are needed with the approach shown in Figure 5.3 to deal with different operating situations:

- If there is no video signal, or the level is very low, automatic focusing should be disabled.

- When video is present and the autofocus system is first activated, focus may be so far off that no part of the image is in focus, meaning there is no high frequency energy signal. This requires that the focus actuator be slowly slewed through its range to find the place where a signal appears. Then the loop can operate normally.

- If a slewing cycle is completed and nothing is found, the system should either shut down or it should switch its filters so that it looks for coarser energy patterns on which to operate. If this still fails, the system must shut down. This raises the question, then, of what stimulus causes it to retry. The best choice is probably to watch for a significant change in video level and then have the focus system retry.

- When focus has been acquired, the small dithering that was necessary for the servo to operate may prove objectionable in the image. At this point, dithering and the servo loop can be shut down. It should continue to watch the focus error signal, however, and reactivate as soon as any change in the focus signal occurs.

Automatic focus systems are tricky and involve many trade-offs with the operating conditions as shown by the list above. The performance of individual cameras should be carefully evaluated to decide whether it will be satisfactory for specific intended applications.

5.5 IMAGE STABILIZATION

Hand-held cameras must be carefully supported to capture pictures without instability because of operator motion or shaking. Images can be stabilized by electronic means, optical means, or by mechanically stabilizing the entire camera. Each of these approaches has advantages and disadvantages and has been adopted in different markets.

Image stabilization requires two separate steps: detection of instability and movement of the image for correction. There are different methods to do each as described below. The systems are usually named for the method used to move the image for correction.

5.5.1 Detection of Camera Motion

Camera motion detection can be accomplished electronically by processing the image from the camera or it can be done mechanically.

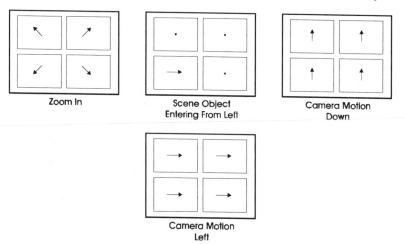

Figure 5.4 *Image stabilization motion vectors.*

5.5.1.1 Electronic Detection of Image Motion

Motion causes changes between adjacent video frames. Of course, the motion could be camera motion or it could be scene motion, and the camera motion could be intentional (such as panning or zooming) or it could be operator instability that should be removed. It is the job of the detector to separate all that out.

The detection process begins by storing one or more frames of video. Comparison of the same areas between frames can yield indicators of motion. It is useful to do this in a number of blocks in the image because camera instability will cause correlation between the motions of different blocks, while the other forms of motion (except for panning) seldon affect all the blocks the same way. A correction signal is derived by calculating the correlation between motion detection in blocks at several locations in the frame.

The motion detection must determine magnitude and direction of the motion in each block, this is the process of generating *motion vectors*. Figure 5.4 shows some typical situations. One approach to motion detection is to select a number of key points in a block and then examine that block in the next frame and try to find the same key points. Motion vectors can be found by comparing the locations of the key points between the two frames. A motion vector for the block can then be determined by combining the motions of all the key points. The calculation of the correlation between the motion vectors has proved to be a good application of fuzzy logic techniques [1].

5.5.1.2 Nonelectronic Detection of Image Motion

Electronic motion detection has the difficult problem of separating camera motion from scene motion. It also is limited by the detail content of the scene and by the scene illumination. These factors ultimately limit the performance of a stabilization system that uses electronic detection. The solution would be to detect the camera motion directly, without

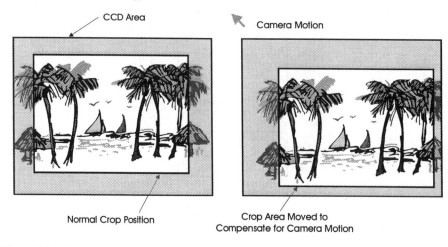

Figure 5.5 *Electronic correction of camera motion by cropping the imager output.*

using the image. This can be done mechanically, using acceleration sensors to sense the rate of change of camera motion.

5.5.2 Correction of Camera Motion

Having determined the camera motion to be corrected, the correction can be accomplished either electronically or optically.

5.5.2.1 Electronic Correction

Whan the camera is inadvertently moved or shaken, the camera's imager receives the wrong optical input. The only way to correct that electronically is for the CCD to scan a larger optical image that is cropped to the final output. Motion correction is then accomplished by moving the crop rectangle. Since some pixels scanned by the CCD are thrown away in the cropping, there is a small loss of resolution compared to the CCD's full capability but, with current high-resolution CCDs, the loss is acceptable in the consumer market. It is not acceptable for the professional market, although sometimes the benefit of stabilization is worth a trade-off in picture resolution.

Figure 5.5 shows how movable cropping of the output of an imager corrects for camera instability. The full output from the imager is stored in a digital memory and the cropping is defined by the addressing of the memory read function. Since the cropped area is reduced in size compared to the full scan area, it must be digitally enlarged by resampling. This is the same function as electronic zoom (see Section 5.6.4).

5.5.2.2 Optical Correction

Motion correction can also be achieved optically. One approach [2] uses a deformable

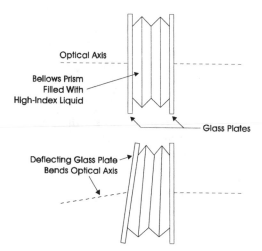

Figure 5.6 *Optical image correction using a deformable prism.*

prism that contains a liquid having a high refractive index. By squeezing one side of the prism, the angle of the prism is changed to shift the optical path in a direction to compensate for the undesired camera motion. This is shown in Figure 5.6.

Optical correction has the advantage that there is no loss of resolution. Disadvantages are that it adds to the optical path length, which may limit the lens choices that can be used with the camera (because of back focal length), and it may be more costly than electronic correction.

5.5.3 Mechanical Camera Stabilization

The image stabilizers described so far are *active* in that they embody electronic or electrical manipulation of the image. Camera stabilization can also be done *passively*, as in an automobile suspension.

Figure 5.7 shows a typical passive mechanical stabilizer. The camera operator guides a platform on which the camera is supported by a freely moving gimbal arm mechanism that is isolated from the operator's motion. The camera's weight is balanced by springs in the arm. The mass of the camera provides its stability and the operator's motion has a minimal effect. As in the automobile suspension, there must be a shock absorber device to limit the relative motion between camera and operator.

Active mechanical stabilization is also possible (see Section 10.3.3).

5.6 IMAGE ZOOM

Optical zoom lenses were discussed in Section 2.3.3.4. Focal length can be varied as much as 70:1. These lenses require sophisticated controls, which are discussed here. Zooming can be done electronically, which is also discussed in this section.

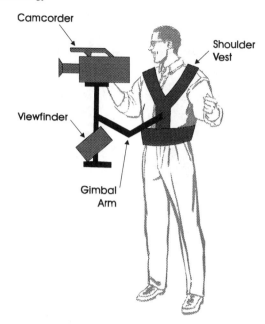

Figure 5.7 *Passive mechanical camera stabilizer.*

5.6.1 Zoom Lens Control

There are three parameters that always must be controlled on zoom lenses: zoom, focus, and aperture. On small lenses as used on hand-held cameras, these may be operated by direct mechanical manipulation. Of course, cameras that have automatic operation for any of the lens features must contain electric motors to control those functions. Larger zoom lenses, as used on studio or field production cameras are impractical to operate directly, simply because the lens is too far from the camera operator. These lenses always have remote controls.

5.6.2 Mechanical Control of Zoom Lens

Zoom lenses are usually designed to be zoomed and focused by rotating rings on the outside of the lens barrel. These deliver the appropriate axial movements to the lens elements through a *barrel-cam* mechanism that converts the rotation to a linear motion. Small lenses may be controlled by hand simply by rotating the rings. Motor control is done by gearing a motor to each ring to produce the rotation.

Large lenses, such as those used on studio cameras require remote control simply because they are too far forward for the operator to reach the focus and zoom barrels. One method for handling that is to provide flexible shafts that are coupled to the lens rings in the same way motors would be. The operating end of the flexible shafts is brought to the back of the camera and mounted to the camera positioning handle. Thus, the control is

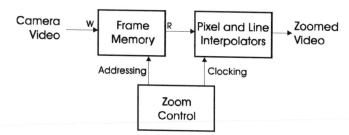

Figure 5.8 Block diagram of electronic zoom feature.

convenient for the operator to use without moving his or her hands from the positioning handles.

5.6.3 Electrical Control of Zoom Lens

More sophisticated zoom lens control is available if electric motors controlled by servomechanisms are provided. This allows control of the lens from a control position away from the camera or it allows automatic control. Features such as variable-speed control, or memory can be provided. The memory feature allows various shots to be established and stored during setup of the scene; during actual shooting, they are called up simply by pressing a button. A controller of this type is called a *shot box*.

5.6.4 Electronic Zoom

In the home market, an electronic zoom feature is often provided to extend the range of zooming beyond the lens range. This feature is especially easy to implement if the camera is already using digital processing. By storing each frame in a digital memory and controlling the memory read function while interpolating pixels (and maybe lines), the picture can be enlarged by any factor. Of course, too much enlargement in this way will bring the pixels themselves into visibility, and picture quality is reduced. It becomes a subjective trade-off between the effect of the zoom and the loss of picture quality. Many users find that it is a valuable trade-off and the feature is popular.

Figure 5.8 is a block diagram of an electronic zoom system. When electronic zoom is activated, it is usually set up to seamlessly come into play when the lens zoom reaches its maximum telephoto position. Thus, the operator feels there is an extended-range of zoom accessible from the one control.

REFERENCES

1.   Egusa, Yo, et al., "An Application of Fuzzy Set Theory for an Electronic Video Camera Image Stabilizer," *IEEE Transactions on Fuzzy Systems*, Vol. 3, No. 3, August 1995.

2.   Sony SteadyShot.™

6

Cameras Used in Systems

Cameras are seldom used by themselves; they work with other cameras, recorders, switchers, monitors, and so forth. These external devices place requirements on the design of cameras that must fit into systems.

6.1 WHAT IS A SYSTEM?

The dictionary defines a system as "a group of interacting, interrelated, or interdependent elements forming a complex whole" [1]. With this definition, a camera itself is a system; however, this chapter is concerned only with systems where a camera is an element rather than the whole system. Features of cameras that relate to their involvement in systems are things such as signal interfaces, remote controls, voice intercom, synchronization, cables, and so on.

6.2 WHY SYSTEMS ARE NEEDED

Systems are required when multiple cameras are used together at the same scene or when all the people involved with the operation are not at the same location. The most common use of multiple cameras is in *live video production*, where the signals from several cameras are combined as they are being shot to produce a complete production. This is common in television for present sporting events, political conventions, and other scheduled events that must be aired while they happen.

On the other hand, production of programs that are recorded for later use is seldom done live. The preferred method for this is what is called the *production-postproduction* style. In this approach, camera shots are done separately and recorded for later assembly into the program. The program assembly is called *postproduction* and it has many advantages in terms of the program quality (*production values*) that can be achieved and in the editing capabilities available for use in the program. Because postproduction is free of the

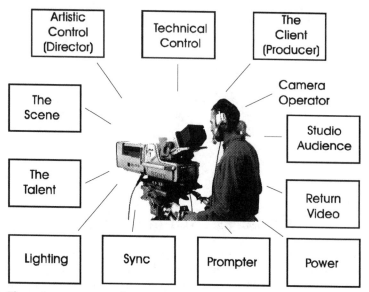

Figure 6.1 *The environment surrounding a camera. (Photograph courtesy of Sony.)*

constrainst of the live scene, more time can be devited to it without adding to talent or on-location costs.

The system surrounding a camera may be thought of as the camera's *environment*. The various elements that may exist in the environment surrounding a camera are shown in Figure 6.1. A home system has very few of these items, but a professional video production studio may contain all of them.

The environmental elements are both mechanical and electrical, and they deal with the people in the studio as well as the technical facilities behind the scenes. The way these external factors affect cameras is the subject of this chapter.

6.3 SYSTEMS ISSUES

This section introduces each of the systems issues and explains its purpose. Later sections of this chapter will discuss the issues in detail.

6.3.1 Signal Interfacing

Most cameras, even camcorders, have some means for delivering video and audio signals to an external display or system. Some cameras also *receive* signals from a system.

6.3.1.1 Audio and Video Signals

Video, and possibly audio signals captured by a camera, must somehow get out of the camera to where they will be edited or viewed. A camcorder achieves this through its tape, but all other cameras must output their signals electronically. This is normally by means of one or more cables, which may be copper or fiber-optic cable. In the case of analog video and audio, signal formats are usually NTSC, PAL, S-Video, or component. With digital cameras, most cameras use serial communication, which has the advantage of only a single cable connection and the ability to multiplex multiple signals on that one cable. The most popular serial digital formats for cameras are SMPTE 259M and Institute of Electrical and Electronics Engineers (IEEE) standard 1394 (see Section 6.5).

6.3.1.2 Return Video

When cameras are used for shooting scenes that are (or will be) combined with other shots, it is useful to provide information at the camera viewfinder to assist the camera operator to properly frame the shot. That can be accomplished by sending video signals from the system for viewing on the camera viewfinder. System cameras typically have provision for receiving video from the system and displaying it on the viewfinder along with the camera's own signal. This is called *return video* capability. A separate video feed to the camera is often provided to send signals to a teleprompter or monitor mounted on or near the camera.

6.3.2 Synchronization

Where signals from multiple cameras are to be switched or mixed in real time, there must be means for *synchronization* of the cameras, so their signals will combine properly. Camcorders that operate without any system connection do not require this because they generate their own timings, which are recorded on the tape. Synchronization with other sources is necessary only when the tape is played back.

The basic synchronization requirement is to coordinate the scanning patterns of all the cameras in the system so that video signals can be combined at a mixing point. Additional requirements exist in systems that use subcarriers or other forms of modulation; the modulation parameters also must be synchronized. A *sync generator* delivers synchronization signals that are sent to all cameras and other units in the system. The sync signal distribution must account for cable delays so that signals are accurately synchronized at the point where they will be brought together (usually at a video switcher).

Modern systems generally use a composite form of synchronization signal that combines horizontal and vertical scanning sync with modulation sync in one signal. In NTSC or PAL systems, this usually is a *black burst signal*, which is a standard composite video signal having sync pulses and color subcarrier burst, but the picture is all black.

In digital systems, synchronization of the cameras themselves could theoretically be avoided by using frame storage at the mixing point to provide buffering from which all signals are read out synchronously (*frame synchronizers*). However, that is usually a more expensive approach than sending synchronizing signals out to the cameras except in the case of truly remote cameras where sending sync may be impractical. Thus, digital systems generally use a synchronization approach similar to analog systems.

6.3.3 Communication

When a camera is part of an extended system, there is need to coordinate the different people involved in the production. That is generally handled by an intercom subsystem in the cameras that provides voice communication between the participants. There may be more than one intercom circuit to support the needs of both the artistic and the technical staffs (see Section 6.8.2).

6.3.4 Remote Control

Most multicamera systems have centralized facilities for technical and artistic control of the production (see Section 6.4.1). That inherently requires a degree of *remote control* of the cameras. Another reason for remote control is to supplement the skill of the camera operator and relieve him or her from all tasks other than the framing, focusing, and moving of the camera. Some cameras are in locations where an operator cannot be present; obviously these require full remote control of all functions including positioning and focusing. Remote control systems are covered in Section 6.7.

6.4 SYSTEM CONFIGURATIONS

The requirements outlined in the previous section are supported by many different system configurations. Several of those will be described in this section.

6.4.1 Studio Production System

A professional video production plant that produces complex video programs for broadcasting or other distribution generally has several studios, each providing for at least three cameras. Each studio has its own control room for artistic control, but technical control may be centralized for all studios.

During production, camera outputs may be separately recorded, or they may be combined in video switching, which is then recorded. A studio also is capable of producing a complete live program when that is necessary. All combinations of these modes should be possible. Recording equipment is also usually centralized for the entire plant.

6.4.1.1 Routing and Assignment

An important requirement in a large production facility is to have the maximum flexibility to deploy the plant's total equipment complement dynamically to the tasks at hand. That is achieved by the use of a *routing switcher*, which allows any camera or any recorder to be assigned to a particular studio. The assignment of equipment can be modified at any time as the workload changes in the studios. This is accomplished with a minimum of physically moving of equipment or changing cable connections.

The plant is designed for a certain maximum number of camera channels that can be used in each studio. Camera cabling is built into the plant for this, but camera heads are connected in each studio only as needed. Thus, a smaller complement of cameras can be dynamically distributed between the studios than if each studio had a fixed number installed.

A full block diagram of a routing switcher system would take several pages and is not necessary to understand the impact of this architecture on camera design. Figure 6.2 is a simplified diagram showing the important features. This plant has a capability of up to twelve cameras operating at once in several studios (probably four). Eight video recorders are also available in the system, to be assigned wherever they are needed.

Studio A in the figure is set up to have three of its four possible cameras activated. That is accomplished by plugging in the camera heads to the connectors in the studio and activating them by assigning those cameras to studio A, using the assignment panel shown in the technical area. The act of assigning causes the camera video signals to be routed to the switcher in the Studio A control room and their control activated in the technical area. Recorders for use in studio A are also assigned, which makes their controls available in the studio A control room. Alphanumeric displays at each camera control and monitor position change to show the current assignment of those units.

The routing switcher takes care of connecting camera video signals, control, and intercom cables according to the routing plan set up at the assignment panel. The implication to the cameras is that their capabilities be available to the routing switcher. This is usually provided by separating the camera system into two parts—a camera head and a *camera control unit* (CCU). The camera heads move from studio to studio but the CCUs are all collected together at the technical central area. Video, audio, control, and intercom are available at the CCU and are permanently connected to appropriate inputs on the routing switcher.

6.4.1.2 Two-Unit Camera Architecture

The separation of a camera into a camera "head" unit and a camera "control" unit is an architecture that was mandatory in the early days of television simply because the hardware for a camera was too large and heavy to handle as a single unit that could be moved and positioned by a single camera operator. Today, the size and weight of hardware is no longer an issue and self-contained high-performance cameras are available in packages weighing as little as 10 pounds, and may even include a video recorder. However, there are other reasons why two-piece cameras still exist today at the high end of the market. These are:

1. Considering the number of connections required between a camera and the system it is used in (see Section 6.3.1), it is impractical to bring all these directly to cameras that are moving around in a studio or field production site. Some form of multiplexing is required to simplify the cabling to the cameras and to make the camera's timing in the system independent of the camera cable's length and delay. This inherently requires a base unit that accepts the system connections and interfaces them to a single cable going out to the camera. Thus, two-unit construction is inherent in simplifying the camera cable connection. This is met by a camera

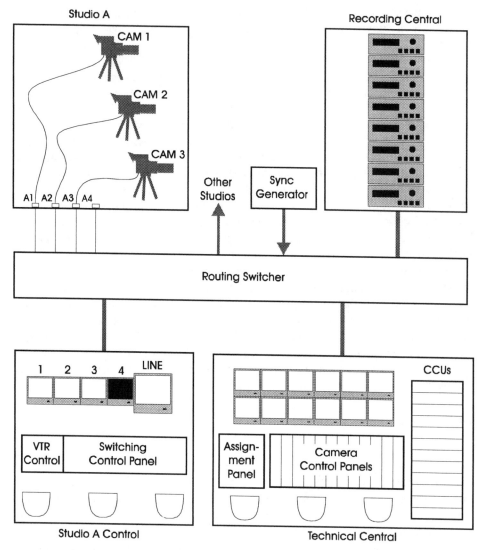

Figure 6.2 *Block diagram of the camera portion of a studio production system.*

"head" and a system interface unit or base station; connected by a camera cable that is simple, low cost, and can be extended to long lengths.

2. Centralization of technical operations requires that the camera interface units be rack-mounted in a technical control area. There are no size and weight limitations at that location, and camera designers have made the choice to move certain of the camera's processing to the interface unit, to simplify interfacing requirements and reduce system cost. Since an important part of the base station unit involves con-

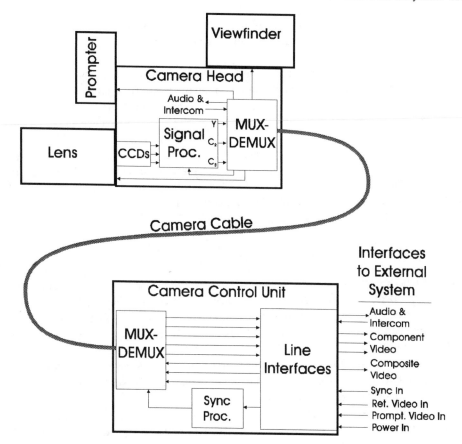

Figure 6.3 *Camera partitioning into camera head and camera control unit.*

trol, it is natural to call it a camera *control* unit (CCU).

3. The camera head is subject to rough handling as it is moved around, especially in field use, and it may also experience environmental extremes. That means that it should be as simple and reliable as possible, leading to the desire to minimize electronic functions in the head. Since the camera head sometimes may be in an inaccessible location, the need for service or maintenance of that unit should be minimized.

A typical architecture for separation of a camera system between head and CCU is shown in Figure 6.3. The camera head and CCU each contain multiplexing-demultiplexing units to allow the camera cable to be a simple single cable. Camera cables are discussed in Section 6.6. The camera head contains video signal processing up to the production of component signals; YC_RC_B format is shown in the figure, but RGB may also be used. The multiplexing may be analog or digital; it supports transmission of the three-component video signals, two return video signals for viewfinder and teleprompter, two-way audio for program and intercom, and remote control of all processing and lens functions in the camera.

Camera power is also delivered through the camera cable. Typically, in addition to powering the camera circuits, a small amount of ac power is provided to an outlet on the camera for the use of small tools, oscilloscopes, lights, heating blankets, monitors, or other accessories. In field situations, the camera may be located in a place where no other source of power is available.

The camera control unit also receives a sync signal from the external system. Processing in the CCU provides proper timing of sync going into the camera cable so the video signals coming back from the camera have the correct timing for integration into the external system. This may involve a feedback loop that automatically compensates for the delay of different lengths of camera cable.

6.4.2 Outside Broadcast System

Field production is sometimes referred to as *outside broadcast* (OB) production. This includes on-location production for dramatic or documentary videos, sports event pickups for broadcasting, political conventions, or other special events that require a system of multiple cameras. Much OB production is done for news happenings, but these normally occur unpredictably, which precludes the setting up of any system. News shooting is usually done with individual camcorders and program integration is done by playing back tapes at a central location or studio. Signals may also be sent back to the central location by satellite or microwave links.

Large OB systems exist at championship golf courses, baseball or football stadiums, basketball courts, or other venues where events occur often enough to justify the building of a permanent on-location system. Important scheduled events, such as the Olympic Games or a political convention, require a large system to be built ad hoc; the system is set up for the event and torn down afterwards. All of these applications require a combination of live and recorded production.

Program producers desire to have the same capabilities at these field production sites that they have in a studio. The systems can get very complex. However, field production sites are seldom in continuous use, so the system design often includes provision for easy removal of expensive equipment units for use at other sites. The permanent installation consists of only those parts that are difficult to reinstall and remove every time the site has to be used.

Figure 6.4 is a simplified diagram of an installation at a championship golf course. The system consists of permanent installation of camera cable around the course, buried underground and terminating at connection boxes or ports at each hole of the course. Camera heads are plugged in at each hole as the action of the tournament moves there. They can be easily moved during the game, so it is not necessary to have cameras installed at each hole all the time. This system provides a tower-mounted camera behind the green of each hole (the tower is permanently installed), a roving camera at each green,

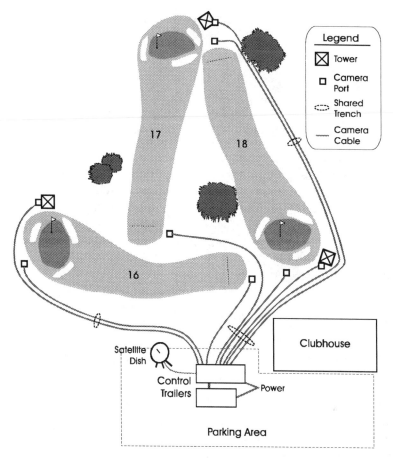

Figure 6.4 *Partial diagram of the video system setup at a championship golf course.*

and a tee camera for each hole where the tee is located such that it cannot be covered by the other cameras.

All the camera cables terminate at a central location, usually a place accessible by large vehicles. The control and switching facilities are located in vans or trailers that are brought in only during an event. These can be moved to another use when the event is over. The central location for this should include appropriate large-vehicle access and parking space and power availability for the vans or trailers. This application shows the need for inexpensive camera cable that can be used in long and variable lengths.

The control facilities in the trailers provide as nearly as possible the same capabilites that are available in a studio system. Because there may be many more camera cables going out to locations on the course than there are physical cameras, a routing system may be included to dynamically connect and reconnect cameras to the control positions during a tournament. That is a convenience; the system could be configured by plugging camera

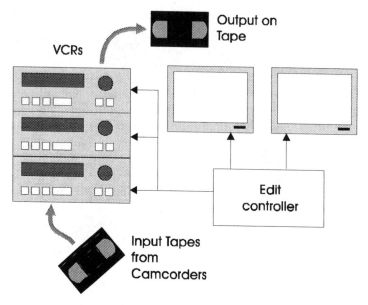

Figure 6.5 *Videotape-based production system.*

cables, but that would be awkward to do during actual shooting.

6.4.3 Videotape-Based Production Systems

System requirements for semi-professional or home production are vastly simpler than the large professional production systems just described. The main reason is that the interconnections between cameras (which are camcorders) and the central system is made via videotape instead of camera cables. There is no centralized real-time control of the cameras; each camera operates by itself under local control of its own operator. Such equipment is much less expensive, but there is no capability for real-time assembly of signals from multiple cameras as is possible in professional production systems.

There may be a need for central artistic control of the shooting. For example, if two or more camcorders are simultaneously shooting at a wedding, the operators may need to communicate to be sure that all shot possibilities get covered. Since there is no cable connection between the operators, intercom has to either be by word of mouth or by radio. There is not much need for built-in intercom in a camcorder.

Figure 6.5 shows a typical system using videotape for bringing in shots from the shooting as in a conventional production-postproduction operation. The cameras operate independently, creating their own tapes, which are brought to the central editing setup. Editing may be done directly with videotape or, the signals may be captured in a computer for *nonlinear editing* [2]. Certain camcorder features are valuable in this type of operation:

1. The camcorder should have *time code* capability. This means that the camcorder places a standard code on all its tapes, separate from the video but in synchronism with it, that indicates the time base of shooting on a frame-by-frame basis. Time code is used during editing to precisely locate the exact frames where program assembly events take place. The most popular time code is the SMPTE Time Code, defined in SMPTE Standard 12M.[3] Home camcorders generally do not have time code and editing must be done on a *tape time* basis, which means editing points are based on physical locations along the tape. That is much less precise than time code.

2. Most camcorders have built-in microphones for sound pickup. That is convenient but, in many cases, it does not produce high quality audio. The main reason is that a camera-mounted microphone is likely to be too far from the sound sources in the scene and is subject to excessive pickup of ambient sound. A camcorder should have provision for use of an external microphone.

3. Although the principal use of a camcorder is to create videotapes, there may still be need for video output directly from the camcorder. If the camcorder is physically brought to the editing station, it can play its own tapes into the editing system, which reduces by one the number of VCRs required for editing. This type of use also requires that the camcorder have a remote control interface so its operation can be controlled from the edit controller. This form of operation is especially important to home systems, where the cost of two or even three VCRs for editing may be prohibitive.

6.4.3.1 In-Camera Editing

Home-use camcorders generally provide some features for editing the videotape directly in the camera without external equipment (see Section 8.5.2). Separate shots can be recorded one after another (this is called *add-on editing*) with certain types of dynamic transitions between them, or a particular shot may be rerecorded without causing interruption in the synchronization of the tape (this is called *insert editing*). When properly used, these features can produce a final tape of a program that appears to have been done in one continuous session. However, there are many limitations compared to true postproduction-style editing using separate equipment.

A camcorder in the field has only one input—the optical scene. If a scene is judged unsatisfactory upon immediate replay, it must be recorded again by having the scene rerun before the camera and repeat an add-on edit at exactly the same place on the tape. The previous take is lost. (In shooting for postproduction, all takes are generally kept on the tape, allowing the possibility of a later decision to choose a different take or part of a different take.) This problem becomes worse in the case of insert editing that rerecords a scene shot earlier on the tape—the scene must rerun and it must be exactly the same length as the previous one it is recording over. These are artistic limitations of in-camera editing.

When a scene that has been shot earlier is declared to be too long, there is no way to correct that except by rerecording the too-long scene *and all the scenes after it*. Although it may be useful to a home videographer, in-camera editing features are not very useful in any type of quality production. It is generally best to record all takes in the field and perform editing back at the home base where an editing system of much greater capability can be provided.

6.5 DIGITAL INTERFACE STANDARDS

The signal interfaces of analog cameras have been standardized for many years. Home equipment generally make both video and audio connections using "RCA phono" connectors, which are push-on two-contact connectors. In professional equipment, however, video signals use 75-ohm coaxial cables with BNC connectors, which are bayonet-type coaxial cable connectors that provide higher reliability. Video signal standards for composite systems like NTSC or PAL define the signal properties and signal levels.

Similarly, professional audio uses audio cables with XLR connectors, which are multiconductor connectors (up to four with shield) that have a reliable locking mechanism. However, all this changes in the digital era—audio and video signal formats, cables, and connectors may be different. This section discusses two of the most popular digital interface standards.

6.5.1 SMPTE 259M

The SMPTE has defined both parallel and serial digital interfaces for professional applications [3]. The parallel interface standards are SMPTE 125M for 4:2:2 component digital signals and SMPTE 244M for digitized NTSC signals. Parallel interfaces are suitable for short-distance connections such as within a control room, but they are not suitable for long distances because of the complexity of the cabling and the precise requirement for timing match between the individual lines in the cable. Longer distances require serial digital interfacing, which generally uses a single coaxial cable, twisted-pair cable, or fiber-optic cable.

The SMPTE standard for serial digital interfacing is Standard 259M, covering both 4:2:2 component signals at 4:3 or 16:9 aspect ratios and $4f_{SC}$ digitized NTSC or PAL signals. 259M defines a serial bit stream and a signal-level interface; it refers back to 125M and 244M for the details of digital video encoding. 259M is intended for use with 75-ohm coaxial cables, which is an advantage to existing facilities that are converting to digital equipment because existing cables installed in the plant for analog signals can still be used. Depending on the type of cable and equalization used, 259M can support distances up to 1,000 ft with high-performance 75-ohm coax.

The bit rates for SMPTE 259M are 270 Mb/s for 4:2:2-4:3, 360 Mb/s for 4:2:2-16:9, 143 Mb/s for digital NTSC, and 177 Mb/s for digital PAL. These bit rates are based on 10 bps; if 8-bit samples are used, two LSB zero bits are added to the data to pad up to 10 bits. The channel coding is scrambled NRZI [4] and the nominal signal amplitude is 800 mV.

The digital encoding defined in 125M and 244M does not digitize the blanking intervals and does not expand the active lines to fill the blanking intervals. Thus, blanking time is available for adding ID codes, error protection, or ancillary data to the 259M bit stream. An important use of the ancillary data space is for audio and, an additional standard (SMPTE 272M) defines transmission of two or more digital audio channels in the ancillary data space of 259M.

Several manufacturers are offering chips or chip sets for 259M transmission, making it easy for camera manufacturers to include that interface in their products. Chips are also available for switching 259M bit streams for routing-switcher applications. 259M is a point-to-point interface that requires terminal equipment at each end of every cable. This limits its flexibility and increases cost, restricting its use to professional cameras. Home and semi-professional cameras are more likely to use the less expensive and more flexible IEEE 1394 standard, described in the next section.

Broadcasters often use home-type digital camcorders for news shooting where the small size and low cost of such cameras is advantageous. For this application, a conversion device is required to convert 1394 to the 259M typically used in the studio.

The protocol of 259M has also been adapted to fiber-optic cable use (SMPTE 297M) and for HDTV transmission (SMPTE 292M). The latter standard supports data rates in the range of 1.3 to 1.5 Gb/s.

6.5.2 IEEE 1394

A much more flexible connection architecture than the unidirectional point-to-point structure of 259M is a two-way bus-oriented approach that allows connection of multiple devices to the same point. That is the internal architecture of PCs and the IEEE 1394 standard [5] offers the same approach for external connections. Unlike the internal bus of a PC, which is parallel, IEEE 1394 is a serial bus, but it can perform the same functions as a parallel bus.

IEEE 1394 uses a four- or six-wire shielded cable. Two twisted-pairs in the cable provide two-way data transmission and two power-distribution wires are optional. (Note that all nodes in an IEEE 1394 network contain active devices and require power at all times the network is in operation, even if their host device is powered-off.) In the 259M architecture, communication over a cable is only possible between the two units connected by the cable but, with 1394's bus structure, communication is possible between any two units anywhere on the bus. Broadcast transmission to all units is also available.

A major target application for IEEE 1394 is digital audio and video distribution within the home. Cable, satellite, or off-air receivers can send signals to displays (monitors) throughout the home. VCRs, DVD players, digital video cameras, or personal computers can also join the bus and exchange signals with the other sources or displays. This is sometimes called a home audiovisual network and it is in many people's vision of the future of consumer electronics. Such a network must be capable of transmitting and receiving multiple video and audio signals while at the same time distributing command, control, and other types of data. IEEE 1394 can do this all at a cost within the range of consumer electronics.

Since the standard was published in 1995, many companies have endorsed it and are beginning to incorporate it into products. To support this activity and to foster more development and new applications, the *1394 Trade Association* [6] has been formed and is very active.

6.5.2.1 Features of IEEE 1394

The IEEE 1394 serial bus standard offers a "backplane" version suitable for communication between modules in hard-wired equipment and "cable" versions for wired connection of separate units. Only the cable versions are discussed here. Some of the features of the IEEE 1394 cable versions are:

- Data rates are specific multiples of 24.576 Mb/s: ×4, ×8, and ×16. These rates are close to 100 Mb/s, 200Mb/s, and 400 Mb/s and the modes are therefore called S100, S200, and S400, respectively. Within the limits of the physical hardware provided, the bus supports intermixing of devices operating at any of the data rates.

- The standard IEEE 1394 cable has a maximum length of 4.5m (15 ft) between nodes when supporting modes up to S400. Up to 16 cables can be in series on a single network, giving a maximum network length of 72m. If only the slower modes are used, cables can be somewhat longer. These lengths are generally sufficient for interconnecting units in the same room or on the same desktop, but they are not long enough to fulfil the whole-house objective stated above. Work is ongoing in the standards groups to achieve longer cable lengths.

- Real-time audio and video require signal delivery that is continuous, on-time, and free from interruptions. Of course, these problems can be accommodated in any network having sufficient data rate, and the receiving devices contain enough buffering of data to ride through delays or interruptions. However, that is expensive and not 100% reliable. *Just-in-time delivery* is much better. IEEE 1394 provides an *isochronous* mode that guarantees just-in-time delivery of multiple audio and video signals without giving up the normal asynchronous mode that is standard in most digital networks.

- IEEE 1394 is a *peer-to-peer network*, meaning that all nodes have identical capabilities. No master control is required; all bus control is achieved by cooperative behavior among identical nodes. All nodes contain a standard set of memory registers that hold information about the network as a whole. These are initialized during the common bus initialization procedure that occurs on startup or whenever a physical change occurs in the network structure.

- A 1394 network can be built in any topology: star, tree, daisy chain, and so forth; the only restriction is that there be no loops. A maximum of 63 total nodes can be in one network, within the restriction given above of no more than 16 in series. The network is self-configuring; when devices are added or removed during operation, the network automatically readjusts itself. This ability to accommodate

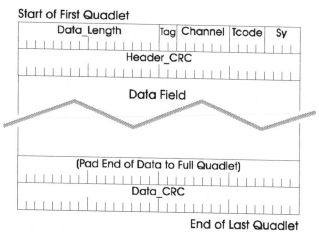

Figure 6.6 *IEEE 1394 isochronous packet structure.*

physical changes during operation is called *hot-plugging.*

These features are just the beginning. The 1394 Trade Association is sponsoring many new developments and this technology will be used extensively in both home eletronics and personal computers.

It is not necessary to the objectives of this book to cover the operation of IEEE 1394 in detail. Several papers on that subject are available from the 1394 Trade Association Web site [6]. However, some of the features will be expanded here to help the reader appreciate the power of this standard and its application to video cameras.

6.5.2.2 Operation of IEEE 1394 Networks

Data transfer on IEEE 1394 is by means of variable-length packets (see Section 1.8.2.3). The packet header consists of two 32-bit words (called *quadlets*), that are defined as shown in Figure 6.6 for an isochronous packet. The minimum packet is a header only—two quadlets; the maximum packet based on the16-bit Data_Length field could be up to 65,535 quadlets of data, but that would span many bus cycles and, therefore, is not practical for isochronous operation. IEEE 1394 recognizes this and restricts maximum isochronous packet length to a maximum of 256 quadlets at the S100 rate. In practice, isochronous packet lengths must be determined by partitioning the input data rate into 125-µs blocks (see the next paragraph). This generally leads to smaller packets than the IEEE 1394 limit. For example, an input data rate of 20 Mb/s requires transmission of at least $20 \times 10^6 / 8,000 = 2,500$ data bits in each isochronous packet. That is about 80 quadlets per packet.

The isochronous mode is most applicable to the transmission of video signals from a camera to other users; it allows multiple video signals to be transmitted in real-time, without interference, interruption, or excessive delay. This mode sets up cyclical opera-

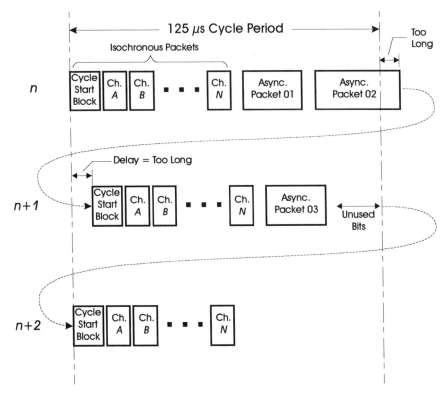

Figure 6.7 IEEE 1394 isochronous operation.

tion of the bus, with a 125 µs ± 12.5 ns period (corresponding to a rate of 8,000 Hz). The first node requesting isochronous operation performs the setup process, and begins transmitting a special sync block called the *cycle start* block, at 125 µs intervals. Because there may be asynchronous transfers going on, the cycle start block may not always acquire the bus on time. That causes a delay in starting the cycle; but the amount of any delay is encoded within the cycle start block, allowing users of data to correct for any them if necessary. Figure 6.7 is a diagram showing the operation of the bus during isochronous transfers of several channels (*A, B, . . . N*). Notice that all isochronous transfers occur at the beginning of a 125-µs bus cycle; time remaining at the end of a cycle is available for asynchronous transfers. The delay problem referred to above occurs when an asynchronous transfer begins in one bus cycle and does not complete before the next bus cycle is due, as shown between cycle *n* and cycle *n+1* in the figure.

The bus speed (S100, S200, S400) has to be high enough to support the sum of data rates for as many channels as will operate simultaneously plus an additional factor to account for header overhead and the time gaps that are inherent in this type of networking. A typical allowance for this is about 25% of the total data rate.

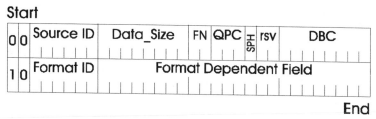

Figure 6.8 *The CIP header structure.*

6.5.2.3 Using IEEE 1394 for A/V

Although the IEE 1394 packet header provides for general isochronous data transfer, more is needed to support audio or video data transfer. The 1394 Trade Association has specified a second level of protocol for this purpose. An additional two-quadlet header is placed at the start of a data block and is called the *Common Isochronous Packet* (CIP) header; its layout is shown in Figure 6.8. The CIP signals its existence in the 1394 packet header by setting the 1394 header's tag field to 01.

Since A/V data streams such as MPEG-2 are already packetized at their source, an important task of the CIP protocol is to translate from one packet structure to IEEE 1394 and back again. Incoming packets are broken up, if necessary, and recreated into their original format and timing at the receiving device.

The fields of the CIP header are defined as follows:

Source ID—identifies the transmitting source.

Data_Size—8 bits define a packet size up to 256 quadlets. That is the maximum for S100; at S200 or S400, multiple data blocks can be included in one packet.

FN—is the fraction number, which defines the number (if any) by which the source packet is divided.

QPC—is the quadlet padding count, which is used when the FN indicates a divisor.

SPH—1 in this bit indicates the source packet has its own header.

rsv—reserved for future use.

DBC—is the data block count, a continuity count that is incremented on every data block. This helps keep track of the data block sequence.

Format ID—This is a code that indicates the format of the data in the packet, such as DVCR, MPEG-2, and so forth.

Format Dependent Field—This field is defined specifically for each format ID.

The content of the fields indicates the flexibility of the Common Isochronous Packet specification for transmission of almost any present or future digital audio/video signals streams.

6.6 CAMERA CABLES

As described in Section 6.4.1.2, a two-unit camera requires a special cable between the units, the *camera cable*. This section discusses the considerations for camera cable selection and some of the approaches that are in use.

6.6.1 Camera Cable Requirements

Many signals must pass between camera head and camera control. Some cameras use multiconductor cables to handle the signals but multiplexing all the signals onto one or a few conductors has become eminently practical as the performance and cost of multiplexing improves. Multicore camera cables are no longer used in new installations. The following discussion applies to both multiconductor and multiplexed cables.

6.6.1.1 Cable Signal Performance

Bandwidth, SNR, and interference susceptibility are the primary signal performance considerations involved with cables. Each of these factors contributes to limiting the maximum usable cable length for a camera. Thus, the length requirement for the camera must be established before considering camera cable performance. For cameras to be used indoors in a studio with all CCUs located in a nearby control room, camera cables seldom need to be longer than about 100 ft. If the cameras are used in a multistudio plant with a central routing and control facility as shown in Figure 6.2, cable lengths can become much longer, depending on the physical size of the plant. Lengths up to 1,000 ft are often required in such installations. Finally, cameras to be used outdoors in a sporting venue such as golf, football, or Olympic games, the camera cable length requirement can grow up to a mile or more. Cable system design varies greatly for these different length requirements.

For baseband analog video signals, the attenuation of coax cables versus frequency (see Section 6.6.2) must be *equalized* to maintain uniform response through the cable. That is simplified somewhat by modulating the video onto a high-frequency carrier, although some equalization is still necessary.

For digital signals, equalization is also a factor in obtaining the highest bit rate on a given length of cable. Digital signals are especially suited to multiplexing, using time division multiplex, packetizing, or any of the numerous other digital multiplexing techniques.

Multiconductor cables are practical only for shorter lengths of cable. In the early days, cameras were designed using multiconductor cables having 100 or more wires and intended for lengths of 1,000 ft or more. Because of cable cost, such lengths of multiconductor cable are ridiculous today and multiplexing is much more economical, convenient, and reliable.

6.6.1.2 Cable Durability

Camera cables are subject to much abuse, even in the controlled environment of a studio. They are dragged along the floor, run over by pedestals and dollies, pulled or stretched, twisted, or even knotted. In the field it may be even worse. Cables are subject to water or other liquids, they are buried underground, run over by heavy vehicles, sometimes even chopped up by construction tools. Mechanical design to withstand these stresses is very important.

Although a cable could be made arbitrarily strong by encasing it in a metal sheath, that trades off flexibility and weight of the cable; it is generally not done with camera cables. Most camera cables use a construction that encases the signal wires in a plastic or rubber sheath, which provides most of the mechanical strength and protection. Metal "messenger" wires may be contained within the cable pack to add tensile strength and take some of the mechanical load away from the signal conductors. In the case of simple cables such as coax and triax, the shielding layer provides tensile strength.

Conductor wires may be either solid metal (usually copper) or *stranded*, where a conductor of a given size is made up of a number of smaller wires twisted together. Stranding allows large conductors to be more flexible and to withstand more flexing than solid construction. Excessive flexing of a solid conductor will cause metal fatigue and eventual failure (breaking).

6.6.1.3 Cable Cost

Camera cables are expensive and their cost can be very significant in the total cost of an installation. When cable lengths of 1,000 ft or more are involved, the camera cable may cost nearly as much as the camera electronics. Thus, it is important to plan installations to use as little camera cable as possible and the cost of that cabling should be carefully calculated and budgeted.

Cost is another factor in favor of multiplexed camera cables. Coax cable costs much less than multiconductor camera cables. At lengths over more than a few hundred feet, a multiplexed-cable camera system (including cables) may cost less than a multiconductor-cable camera system. At the same time, the advantages of a simple cable for mechanical reliability, smaller size, and lower weight are also gained.

6.6.1.4 Cable Connectors

With high-quality cable construction, the weakest link in any cable setup will be the connectors at each end of the cable. Connectors must serve the conflicting mechanical and electrical requirements of physically capturing the end of the cable so it cannot be pulled out, making the electrical connections, yet still be able to be easily and conveniently removed when the system is taken apart or cables are changed. Also, the most likely place for failure of the cable itself is where it enters a connector. This is a point of high stress because the cable can be flexed there but the connector end is held rigidly.

Often, additional semi-flexible protection (*stress relief*) is added over the cable at a

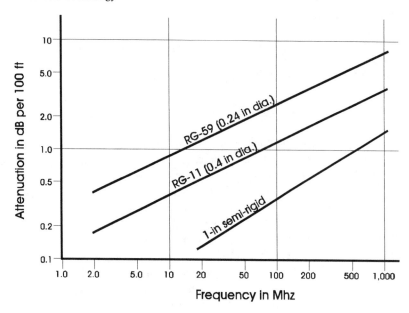

Figure 6.9 *Attenuation versus frequency for typical coaxial cables.*

connector to limit the amount of flexing that can occur there. This is also a trade-off because it causes a stiff section of cable near the camera where it may interfere with operation of the camera. A good camera design will consider this requirement carefully and locate the camera cable connector at the camera so that good cable stress relief can be applied without limiting the operation of the camera.

6.6.2 Coaxial Cable

In analog systems, *coaxial cable* (coax) is the primary medium for transmission of video [7]. It is inexpensive, small in size, and easily handled. Flexible coax comes in various sizes; ranging in diameter from about 0.1 to 0.5 in using a central conductor of solid or stranded wire surrounded by an insulating material (typically a plastic foam material in modern cables), which is covered by a woven shield and a plastic sheath. Semi-rigid coax, which is primarily used for cable TV distribution, comes in sizes from 0.5 to 1 in. It is constructed with a solid inner conductor, foam insulation, and a solid metallic outer conductor, which gives good mechanical strength but limited bending ability.

The principal performance parameters of coax are the *characteristic impedance* and the attenuation versus frequency characteristic. Characteristic impedance is determined by the ratio of diameter of outer to inner conductors and the dielectric constant of the insulating material. Typical impedances for video cables are 50 and 75 ohms. These are nominal figures and there is typically some variation with frequency.

Attenuation versus frequency curves for typical coaxial cables are shown in Figure

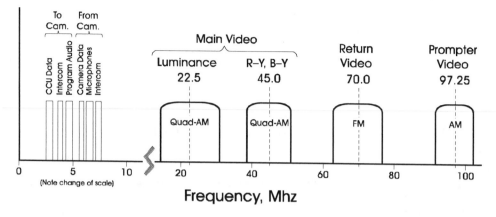

Figure 6.10 *Frequency plan for analog triax multiplexing, from Sony BVP-700/750.*

6.9. Attenuation characteristics depend on cable diameter, construction details, and the type of dielectric material. However, diameter is the strongest determinant; the other factors have comparably small effects.

6.6.3 Triax Cable

The primary medium for multiplexed camera cables is *triax*, which is double-shielded coaxial cable. The center conductor and inner shield function as normal coax to carry the multiplexed signal stream. The outer shield, which is insulated from the inner shield in the cable, provides a system ground connection and an additional layer of shielding for higher-frequency interference. Some triax systems support cable lengths of a mile or more. This section discusses multiplexing methods for triax.

6.6.3.1 Triax Analog Multiplexing

Triax systems with analog multiplexing assign different sections of the cable's bandwidth to the various video, audio, and control signals, a technique called *frequency-division multiplexing* (FDM). A frequency plan for a typical modern high-performance analog triax system is shown in Figure 6.10, specifically the Sony BVP-700/750 system [8]. All signals are modulated on carriers placed according to the plan in the figure. In general, the SNR performance of a modulated signal will be poorer the higher the carrier frequency because of greater attenuation of high frequencies by the triax. Thus, the most demanding channels have lower carrier frequencies and channels such as return video are given the higher-frequency channels. Even so, there is not enough bandwidth for full RGB component video transmission in a system designed for the longest cable lengths.

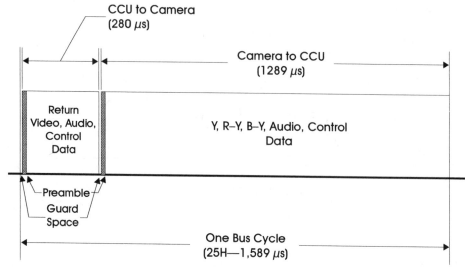

Figure 6.11 *Serial bit stream multiplexing of digital triax camera system. (From Hitachi [9].)*

Four video channels are provided, as follows:

- The luminance channel has a bandwidth of 6.7 MHz, and is double-sideband quadrature amplitude-modulated on a 22.5 MHz carrier. This bandwidth is sufficient for either 4:3 or 16:9 aspect ratio cameras at 525 or 625 lines interlaced, and gives a SNR up to 63 dB. A second signal is transmitted on the quadrature channel for displaying the operation of the camera's skin tone processing on a remote monitor (see Section 4.5.3). That is a special feature of the Sony BVP-700/750 cameras.

- The color-difference channel uses the same modulation as the luminance but on a 45 MHz carrier. The two phases of quadrature are used to separately transmit R − Y and B − Y color difference signals. A bandwidth of 6 MHz is available for each.

- A return video channel has a carrier frequency of 70 MHz and is frequency modulated, giving a bandwidth of 5 MHz.

- A bidirectional teleprompter or auxiliary monitor channel has an amplitude-modulated carrier at 97.25 MHz that gives a bandwidth of 5.5 MHz.

The audio and data channels are placed at the low end of the spectrum on the triax, using multiple frequency-modulated carriers as shown in Figure 6.10. Power is distributed to the camera head also via the triax as a low-frequency AC signal.

Depending on the cable used, this system supports cables lengths up to 2000 m (~1.2 mi) with the return video capability. Sony also offers an optional version that trades off a little bandwidth but can achieve operation at 3000 m cable lengths. In general, analog triax systems are subject to loss of performance with increasing cable length because of the need for greater equalization that introduces noise and because the problems of electromagnetic interference is worse with longer cables.

6.6.3.2 Triax Digital Multiplexing

In a digital camera system, the triax multiplexing also ought to be digital. That can be accomplished at high quality using between 360 to 400 Mb/s serial digital transmission on the triax. A design by Hitachi [9] is described. Digital multiplexing solves the problem of deterioration of performance with cable length but it requires some compromises to achieve long maximum length. Because of digital's consistent performance, these initial compromises exist at all cable lengths.

Figure 6.11 shows the Hitachi system architecture. The serial data rate is 360 Mb/s, and *time-division multiplexing* (TDM) is used to accomplish the bidirectional communication required. The structure of the TDM is shown in Figure 6.11. Since this system is designed for either NTSC or PAL scanning rates, a basic TDM block size corresponding to the time of 25 scan lines (25H) was chosen. That has a period of 1,589 μs. The bulk of that time (1,289 μs) is devoted to the downstream transmission of the main video, program audio, and data. Return video, intercom audio, and data is transmitted upstream in the remaining 280 μs of the 25H period.

Some compression is required on both the downstream and upstream video channels to fit everything into the total data rate. The allocation of most of the transmission time to the main video allows that to be transmitted using minimal compression—the RGB 10-bit digital signals from the camera are digitally converted to Y, R − Y, and B − Y. By not digitizing video information from the blanking intervals and expanding the video data to fill that time, it can be made to fit into the 1,289 μs period at the 360 Mb/s rate. This compression is essentially lossless.

The return video channels can tolerate greater compression and motion-JPEG (see Section 1.8.2.4) was chosen to fit them into the 280 μs period allocated for upstream transmission.

The downstream and upstream data packages must be separated by a time gap that accounts for the maximum cable delay to prevent collision of the bidirectional data packages at one or the other end of the cable. A value of 10 μs is enough for cable lengths up to 1,000m. Each data package begins with a short preamble signal that allows the receiving phase-lock loops to synchronize before any real data comes along. Following the preamble, a header identifies the packet. Most of the packet content is video data, but audio and computer data are interleaved within the video. Audio and computer data account for about 12% of the upstream data but only about 3% of the downstream data.

The design of triax multiplexing still involves a trade-off between signal performance and maximum cable length. This is true for both analog and digital systems. However, digital circuit technology continues to advance and one can expect improvements in digital triax performance in the future.

The camera cable of a camera system is generally considered to be an internal part of the camera and is proprietary to the particular manufacturer and model of camera. Interconnection of camera heads and CCUs of different types is not possible. Therefore, there is not much need for standardization of camera cable signal formats. Especially in the digital world, camera manufacturers will be developing new and better digital formats to keep up with the advancing capability of digital circuit technology. The only constant in this is the cable itself, which can be considered a long-term investment that should not be quickly obsoleted by new circuit technology developments.

6.6.4 Fiber-Optic Cable

A major improvement in the length versus bandwidth equation is offered by fiber-optic cables. Where copper cables (triax) seem to have a practical limit in the range of a few kilometers, fiber-optic cables could potentially extend that to tens of kilometers. Of course, a fiber cannot carry the camera power, but a composite cable including fiber along with power wires is still not too complex, but it is a bit more expensive than triax. The termination of fibers is also no longer a problem. An SMPTE committee is working on a standard for fiber-optic camera cable.

Although several fiber-optic camera systems are on the market, acceptance is so far slow. The reason seems to be purely an economic one. Industry statistics show that about 95% of the cable applications are shorter than 1,000 m, which is readily handled by triax and there is little incentive to replace existing triax installations with fiber. The small number of applications for longer lengths has not made a very large market. However, this will change as HDTV cameras come into use because triax is not practical at the bandwidths required there. HDTV camera cables require digital data rates in the range of 1 to 1.5 Gb/s, which can only be achieved with fiber.

6.6.4.1 A Fiber-Optic Camera System

One of the SDTV fiber-optic camera cable systems on the market is the Panasonic AQ-225 studio/field camera system. A composite cable using two single-mode optical fibers and up to four copper wires provides full-bandwidth serial digital transmission in each direction and camera power delivery. This cable supports power delivery up to 1.5 miles from the base station. With separate camera power supplied at the remote site, fiber-only video, audio, and control transmission can extend as far as 12 miles. In this system, 4:2:2 component video is used and the fibers operate at a data rate of 300 Mb/s in each direction. The same fiber cable will be usable in HDTV systems at data rates up to 1.5 Gb/s (see Section 13.3.5).

6.6.5 Camera Power Issues

The need to send camera power over the camera cable has already been mentioned. This is not as simple as it might seem because of the demands for power at a camera position, which go well beyond the power required by the camera head itself. Cameras often go into positions where there is no local power available so, from the production standpoint, it is desirable for the camera to have "convenience outlets" delivering standard 115 V ac power to devices such as teleprompters, auxiliary monitors, lights, audio equipment, and so forth. That can quickly add up to a load of several hundred watts or more.

The problem is the ohmic loss of power going through the conductors of the camera cable. Conductors have to be kept small to maintain flexibility of the cable. By raising the power voltage on the camera cable, small conductors can carry more wattage but there are still limits in how high the voltage can go for both insulation and safety reasons. The other limit is simply heating of the cable. These factors combine to limit the convenience power of triax systems to a few hundred watts.

Raw ac power cannot be put directly on the camera cable because it has too great a potential for carrying interference. Triax systems convert the input power to a somewhat higher frequency and filter it carefully to prevent passage of any interference from the external power source. At the camera head, the ac power is extracted from the cable and goes to a power converter unit that generates the dc voltages required by the camera electronics. An inverter also generates standard ac power for the convenience outlet; it must include a good protection circuit to prevent overload, which could potentially cause the entire camera to fail.

6.7 CONTROL SYSTEMS

A camera's control system includes its capability for control or adjustment of its operating parameters during operation. These fall into two general classes:

- Artistic controls—these are parameters that control scene-dependent aspects of the camera. They are sometimes called "paint" controls because an operator can use them to "paint" modifications to the picture. In many cameras, especially those intended for the consumer market, paint parameters are automated so the user only has to "point and shoot" with the camera. However, in professional cameras, there are many signal properties that operators may wish to adjust during camera use. Generally, the person at the camera is fully involved with positioning the camera, framing the picture, and focusing. To effectively use paint adjustments, such as brightness, contrast, hue, and so forth, another person must perform those adjustments. That leads directly to the desire for remote control.

- Technical controls—these are parameters that are normally only adjusted during camera setup. With earlier cameras that used tube-type imagers, setup was a major operation and a complex procedure that had to be performed before every camera session. With today's CCD imagers, most of that complex setup procedure is no longer required and "setup" is very simple. However, there is a fuzzy line between setup and operation, and most camera systems provide that almost any adjustment could be made during operation.

The control systems of two camera systems representing the extremes are discussed below: a high-end professional camera system and a good-quality consumer camera.

6.7.1 Professional Control Systems

The control system of the Sony BVP-700/750 cameras is described. These triax cameras have digital remote control of virtually all camera signal processing and lens parameters.

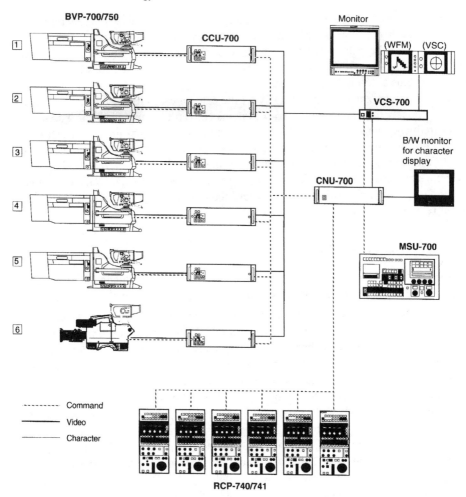

Figure 6.12 *A six-camera control system using Sony BVP-700/750 components. (Courtesy of Sony.)*

The control bit stream is extracted from the triax in the CCU of each camera. An optional remote master setup unit (MSU) can be connected to the CCU for essentially instant access to every setup and operating parameter of any camera in the system. That becomes more effective in a multicamera operation because one MSU can service up to six or more cameras by use of another optional unit, the command network unit (CNU). This unit routes the control channels of all cameras into one MSU, which becomes a central position for all technical operations. This architecture is shown in Figure 6.12.

The MSU is a very elaborate panel intended to give a technical operator essentially instantaneous access to any paramater of any camera. A photograph of its front panel is shown in Figure 6.13. It has a touch-screen display for making selections and rows of buttons for instant access. The system can control cameras individually or it can go into

Figure 6.13 *The Sony MSU-700 Master Setup Unit control panel. (Courtesy of Sony.)*

"All" mode, where the same adjustment is made to all the cameras on the MSU.

There is also a need for simpler remote-control panels (RCPs) that can be deployed on a one-per-camera basis to provide simpler access to the basic operational functions. Sony offers several versions of RCPs to fit different applications. These are interfaced to the system through the CNU.

Monitoring of pictures and waveforms is an integral part of camera control systems. This is handled in the Sony system by another system unit, the Video Channel Selector (VCS) that interfaces to the CNU and provides monitor and waveform monitor switching for six cameras under control from the MSU.

6.7.1.2 Control Data Storage

A very useful feature of a digital remote control system is the ability to store adjustments and return to previous setups at the press of a button. That is provided in the Sony system with an IC memory card, which is a removable card holding nonvolatile memory ICs that store all the setup parameters for a camera. In fact, one card can store up to five "scene files," each of which holds complete paint data for one camera.

The memory card is accessed at the MSU and its contents can be either read into the system or written from the system at this position. The slot at the left of the panel in Figure 6.13 receives the memory card. By means of the memory card it is also possible to enter identical setup information into all cameras or to set up two cameras alike.

6.7.2 Control Systems for Home Cameras

As mentioned above, all setup parameters in home cameras are generally automated so the user only has to point and shoot. However, sophisticated home users or semi-professional users who often buy the same equipment need more than that. Home cameras are designed to operate by themselves and they do not have features to allow them to be used in a multicamera system as described in Section 6.7.1, but there are some important control features even for stand-alone cameras.

- Cameras should have provision for turning off automation and using manual control for automated functions such as focus, iris (exposure), and white balancing. Manual operation of these controls provides a degree of paint capability.

- Home cameras often have built-in features for things that are normally done in postproduction, such as transition effects, titling, or editing. Controls for such features generally have to be simplified because of the limited space available on a small hand-held camera for controls. This leads to them often being awkward to use or placed in inconvenient locations. If these features are important to a user, he or she should make sure they are easy to operate. One consideration that helps is to choose a shoulder-held camera format because that, being larger, provides space for a better control panel. Of course, all these features have the limitations of in-camera editing that were discussed in Section 6.4.3.1.

- Some home cameras provide an infrared remote control unit similar to those used with TV receivers. VCR-type controls are generally provided, such as play, pause, record, stop, and so forth. This is useful when the camera is mounted on a tripod and the camera user wishes to participate in the scene during shooting.

- Home cameras can be used in limited system applications if they provide audio and video outputs. These are useful for external monitoring or copying video to a VCR or computer.

As home and semi-professional users become more sophisticated, it is likely that additional system and control features will appear in these cameras.

6.7.3 Pan and Tilt Control

Cameras that operate in unattended locations require electromechanical mounts to control panning and tilting of the camera. Such devices are available from the simple types used in surveillance cameras up to highly sophisticated servo-controlled units for professional use. Professional camera control systems generally offer a way to expand the control data system of the triax to include control data for auxiliary units such as this. Of course, a really powerful pan-tilt controller may also require more power than the camera can deliver.

6.8 AUDIO

The importance of audio to video pictures is well known and audio is often captured simultaneously with video. Therefore, cameras generally provide features for capture of audio with the video. Triax systems have audio channels to transmit the audio back to the CCU along with the video.

6.8.1 Program Audio

Home cameras generally have built-in microphones for program audio pickup. Some even have stereo microphones. However, the performance of on-camera audio pickup is poor because the camera is usually too far away from the sound source(s) in the scene. The result is a lot of interference from ambient sounds and poor SNR because the desired sounds are low in level. For high-quality sound pickup, it is extremely important to use a microphone external to the camera that can be placed near the sound source. Portable cameras should have jacks for connection of external microphones that feed the sound tracks of the internal recorder in a camcorder or go back to the CCU in a triax professional camera.

6.8.2 Intercom

In multicamera systems, communication is required between the people directing the shoot and the people operating cameras or other equipment at the scene. This is accomplished by the intercom features in the camera system. Several channels of two-way intercom are needed; a simplified schematic of a typical system is shown in Figure 6.14. It provides the following features:

- Two independent intercom circuits are provided for "production" and "engineering." These are "party line circuits," where everyone on a circuit is connected all the time. The production circuit is used by the program director and others concerned with control of the production. During normal production, everyone on the set would use the production circuit. The engineering intercom is used by the technical community for setup or other purely technical purposes. All headsets can choose which intercom circuit to use.

- All headsets can choose to mix program sound into their earpiece. Separate volume controls are provided for the program sound mix.

- All headset microphones can be turned off. This prevents unnecessary background noise from being picked up from unused headset microphones.

- A special "private" mode is provided for the CCU operator to talk privately with the person at the camera. Note that the selector switches at the camera would be forced to the engineering line whenever the CCU operator selected private mode. That is not shown in the diagram.

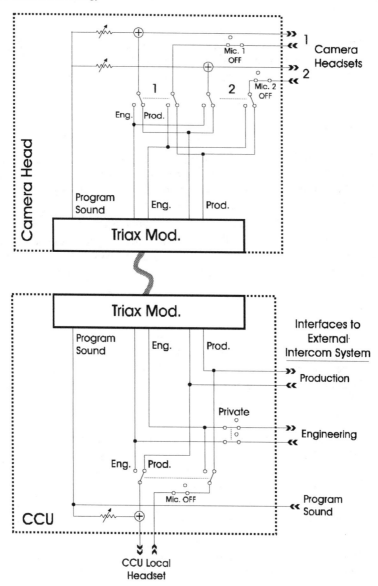

Figure 6.14 *Simplified schematic diagram for a typical camera intercom system.*

Many camera systems provide additional intercom features to expand the flexibility of the basic features described here.

REFERENCES

1. *American Heritage Dictionary*, 3E, SoftKey, International, Cambridge, MA, 1994.

2. Luther, A. C., *Principles of Digital Audio and Video*, Artech House, Norwood, MA, 1997, Chapter 12.

3. SMPTE standards are available from the Society of Motion Picture and Television Engineers, 595 W. Hartsdale Ave., White Plains, NY 10607-1824, or http://www.smpte.org.

4. Luther, A. C., *Principles of Digital Audio and Video*, Artech House, Norwood, MA, 1997, Chapter 7.

5. IEEE standards are available from http://www.stdsbbs.ieee.org.

6. The 1394 Trade Association's Web site is at http://www.a394ta.org.

7. A great deal of information about cables is available at the Belden Wire & Cable Co. web site at http://www.belden.com.

8. Sony Corporation, *BVP-700/750 Product Information Manual*, Sony Brochure BC-00448, Sony Business and Professional Products Group, 3 Paragon Drive, Montvale, NJ 07645.

9. Murata, N. et al., "A Totally Digital Camera System Using Digital Triax Transmission," *SMPTE Journal*, Oct. 1996, pp. 647–652.

7

Telecine Systems

In spite of the advances in video systems that are the subject of this book, motion picture film is still an important ingredient in programming for video, not only archive films but also new production. A special class of equipment is needed to reproduce motion picture film on a video system—the *telecine* system, which consists of a telecine camera, a film transport, and an optical system that images film frames to the camera.

7.1 HISTORY OF FILM IN VIDEO

Before the development of a practical magnetic video recorder in 1956, motion picture film was the only available method for long-term storage of moving pictures or video. All video programs were produced live in real time, which meant that the benefits of postproduction editing were not available to video systems. If a video program had to be stored, it was done with a *kinescope recorder*, a device that had a motion-picture camera in front of a video monitor, resulting in a film of the video program.

Once video recording was developed, the demise of film use in video programs was soon predicted, but in 40 years, that has not happened. Good reasons have caused the two media to coexist, and they still do. The reasons for the continued use of film are:

- Film is a "universal" medium that can be reproduced by any video system. Because film stores images as actual images, its operation is independent of any video signal standard—past, present, or future. Of course, this requires telecine equipment capable of outputting signals for the desired video standard. Films can also be shown in theaters by direct projection.
- The film industry infrastructure for production and postproduction is an extremely capable facility for creation of high-quality programs for eventual reproduction on video systems.
- Film production does not require any knowledge of electronics.

- Massive archives of film shows and movies exist as a resource for future programming.

However, good reproduction of film on a video system is difficult and requires special equipment, called a telecine system. Because of this, film programs are usually transferred to videotape for general use in video systems. Videotape is easier to handle and plays back more consistently in day-to-day video operations. Thus, the telecine has become a *film-to-tape transfer* machine, which is used only by people and companies who specialize in that task. In broadcasting, for example, film-to-tape transfer is performed ahead of air time and on-air programming always runs from videotape, regardless of the source medium of the original material.

Special high-resolution telecine systems have been developed for digital archiving of films. These do not adhere to any existing video standard but, instead, they output digital frames that are stored on digital data tapes. Playback always requires conversion to a video standard for display, but the recordings are just as standard-independent as the original film was. One notable example of such an application was the program to digitize the Fox Movietone News archives consisting of more than 40,000,000 ft of film [1].

7.2 CHARACTERISTICS OF FILM

The image storage capability of film is based on chemical processes that have quite different properties than the electronic image storage means described elsewhere in this book. Those properties cause telecine systems to be quite different from other video imaging systems.

7.2.1 Types of Film

Film consists of a plastic substrate or base material on which one or more light-sensitive chemical layers are placed. The sensitive layers are called the *emulsion.* Images are produced on film by briefly exposing it to an optical image, which causes invisible chemical changes in the layers, creating a *latent image* that can be made visible by the chemical process of *development.* In areas where light strikes the film, development causes a buildup of opaque molecules of silver in the sensitive layer of film, which is otherwise transparent. The degree of opacity varies with light intensity and is usually quantified in terms of the *density* of the film, which is the negative logarithm of the light transmittance through the film.

Because more light produces greater density (less transmittance) in the developed image, this process inherently creates a negative image, where bright areas of the scene appear dark on the film and dark areas of the scene appear bright. To create a positive image that can be viewed normally, it is necessary to image the negative onto another piece of film and develop that, a process called *printing.*

A special class of film, called *reversal* film, produces a positive image directly from the original film, without the need for printing. It requires more complex development and is generally not used in professional motion-picture film production. It is used in photography to create the familiar transparencies, although even there printing is more common, simply because prints are more easily viewed.

Film prints can be either transparent or opaque, corresponding to the familiar motion-picture film or photographic prints on paper. Transparent film is viewed by passing light through it, whereas opaque prints are viewed by illuminating them from the front and depending on the reflection of light from the white substrate to make the densities of the film layer (which is transparent) visible. The following discussion is restricted to transparent viewing only, which is always the form of motion picture film.

The film development process produces an image of only one color, usually black, so the image is monochromatic. To obtain full-color reproduction, color films have three sensitive layers containing dyes corresponding to a set of primary colors. In the development process, the silver images are converted to dye colors to produce a properly colored negative or print.

Motion-picture film comes in various sizes and styles; semi-professional film production usually uses 16-mm (width) film sizes, professional production for video usually uses 35-mm film, and theatrical movie production may use film of 35 mm or greater size. All film has a series of image frames placed along its length, with perforations for film transport along the edges and sound tracks (if used), between the perforations and the frames as shown in Figure 7.1. Film sizes and formats are standardized by the SMPTE [2].

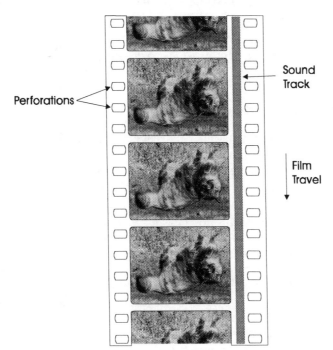

Figure 7.1 *Typical 35-mm negative film, 4:3 aspect ratio with sound.*

7.2.2 Transferring Film to Video

Video transfer from film can be done from either negative or positive film; to a first approximation, the difference is just a matter of inverting the polarity of the video signal. However, the image properties of negative and positive film are very different and the effect on video electronics is much more than simple inversion. Because the rules of accumulated distortions in cascaded analog processes apply to printing a film to obtain a positive image, the image quality will theoretically be better if video transfer is done from an original negative rather than a print.

A telecine system involves the following elements:

1. Means for transporting the film through the system. Depending on the type of pickup system, film transport may have either continuous or intermittent motion (see Section 7.4).
2. Means for illuminating the film. This also depends on the type of pickup system. The illumination may be continuous, as from a lamp; intermittent, as from a strobe light; or scanned, as by a CRT.
3. Means for video pickup. This may be an area-scan CCD, as in a live-pickup video camera; a line-scan CCD; or a photomultiplier tube, as in a flying-spot scanner. Earlier telecine cameras also used photoconductive vacuum-tube imagers.
4. Video signal processing to accommodate the film image characteristics and the frame-rate conversion to video standards.

Obviously, there are many possible combinations of the above elements. However, only a few have seen commercial application and just three will be discussed further below: the line-scan CCD, the area-scan CCD, and the flying-spot scanner.

7.2.3 Transfer Characteristic

Exposure and development of negative film has a typical transfer characteristic shown in Figure 7.2. This is a log-log plot because density is a logarithmic function. The film characteristic has a knee at high density, which provides a "soft" form of highlight compression. At the low-exposure end, the curve tail off into a "toe" that provides the same type of compression for lowlight areas. The slope of the curve in the linear region is the gamma of the film, which is typically about 0.6. If exposure covers the entire region from toe to knee, the exposure dynamic range is about 3.0 log exposure, or 1000:1, corresponding to about 10 stops of camera exposure. Generally, the film should be exposed so that the entire scene brightness range is within the "linear" region, which means that the scene contrast ratio should be less than about 100:1. Since this uses only part of the dynamic range of the film, successful pictures can be made within a range of exposure where the highlights and lowlights remain outside of the highly nonlinear regions. This is known as the film's *exposure latitude*.

The output density range from toe to knee is compressed compared to the input (corresponding to a gamma less than unity); the maximum density range is only about 2.0 or 100:1. That is convenient because it is exactly what the video system requires.

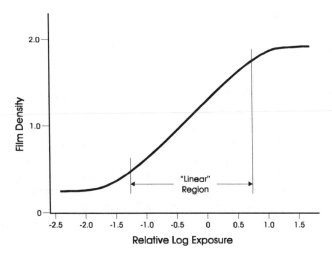

Figure 7.2 *Transfer characteristic curve for a typical negative film.*

The transfer characteristic of a color transparency print is quite different, because print film has a much higher density range and, therefore, gamma. The result is that a print from a negative has the typical curve shown in Figure 7.3. The maximum gamma of this curve is about 1.8, meaning that the 100:1 contrast range of the negative film has been expanded to something like 500:1. Although this gives excellent results for projection viewing in a dark movie theater, it is a major difficulty facing a telecine system because a video system cannot reproduce much more than 100:1 contrast ratio.

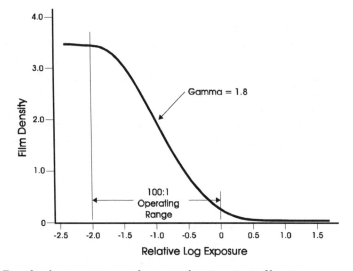

Figure 7.3 *Transfer characteristic curve for a typical motion-picture film print.*

7.2.4 Color Properties

Three-layer color film uses red, green, and blue dye for the taking filters. When the film is developed, the silver molecules in each layer are replaced by a dye of the corresponding subtractive primary color, since viewing of the negative by transmission is subtractive. If a print is made from the negative, the dye colors in the print are the same as in the negative because it also is viewed subtractively.

Being subtractive, a film system produces white by the absence of dyes in the color layers, whereas the additive video system produces white by maximizing the light intensity in each color channel. This causes a fundamental difference between the systems in the function relating color intensity (saturation) to brightness (luminance). In a film system, the most intense colors are achievable when the dye densities are high, corresponding to the dark areas of the scene. A video system is opposite—the most intense colors are achievable when the color channel signals are near maximum, corresponding to the highlights of the scene. Viewing a projected film image with a normal video camera shows washed-out colors and loss of gray scale details in the dark areas of the picture. A telecine camera must fix this problem by its special signal processing (see Section 7.7).

The color performance of developed film deteriorates with age. The dyes fade with time and especially with exposure to light. Old films lose contrast and color. Also, the three dyes do not age at the same rates, so the color balance of the film also changes with age. Further, the dyes may shrink differentially with time, causing misregistration. Capturing old film to video requires constant attention to color balancing, color correction, and other parameters.

7.2.5 Noise Performance of Film

The chemical materials in the emulsion layers of film are granular and their micro-structure in the picture looks very similar to the random noise of a video picture. Different film stocks have different granularity (grain), and usually there is a trade-off between grain and light sensitivity. Finer-grain films require more light for exposure. In video terms, the grain performance of film corresponds to SNRs in the range from 45 to 55 dB. Because different film systems have different image sizes (16 mm, 35mm, and so forth) but the physical grain size in constant for a given type of film, larger-format films produce better SNR values when transferred to video. Of course, the video pickup involved in film-to-video transfer also contributes noise to the pictures, the same as it does in live video pickup.

7.2.6 Summary of Film Performance on Video

Most writings about film versus video compare the media from the point of view of which is the best choice for a particular production situation. That is not the purpose of this comparison; production decisions are already made and we are faced with obtaining the best reproduction of film on a video system. In this context, the considerations are:

- Gray scale reproduction;
- Colorimetry;
- Transporting the film;
- Frame rates.

The first three of these were discussed above. The last item, frame rates, is the subject of the next section.

7.3 FRAME RATES

Modern motion-picture film is designed to run at 24 frames/s. Some older films ran at 16 frames/s. Either of these rates to too low for direct viewing without flicker. In theatrical projection, each frame is exposed twice on the screen, giving a flicker frame rate of 48 f/s and a motion frame rate of 24 f/s. In terms of the discussion about video frame rates in Section 1.4.4, these rates are low; they are satisfactory in theater projection because of the viewing conditions—viewing is in a darkened room, the screen brightness is not too high, and the audience is not too close to the screen.

Film and video frame rates are different—too different to simply speed up the film to match the video rate. Speeding-up is sometimes done in PAL video systems, where the frame rate is 25 f/s; the 4% speed-up required to run 24 f/s film at 25 f/s is generally not noticeable in the video and may be satisfactory for the audio if the audience is not too concerned about pitch. However, it is an undesirable compromise and modern telecine systems with digital processing have frame rate conversion to get around it.

7.3.1 Frame Rate Conversion

In NTSC video systems, where the frame rate is 30 f/s, the film frame rate must be converted. This section discusses frame rate conversion methods.

7.3.1.1 3:2 Pulldown

Analog 30 f/s interlaced video systems accomplish frame rate conversion by an elegant and simple scheme known as 3:2 pulldown, illustrated in Figure 7.4. This is based on the exact 2/5 relationship between film frames at 24 f/s and interlaced video fields at 60 f/s. Conversion is achieved by exposing one film frame to three successive video fields and the next film frame to two successive video fields. Thus, the name 3:2. Because of the uneven exposure of the film frames relative to the video fields, there is a motion artifact that makes some scene motions reproduce a little jerkily. However, the approach has been widely used and audiences have learned to accept its appearance. Figure 7.5 shows what happens to a moving object with 3:2 processing. The film captures the object at its actual position every 24th second but, because the video system repeats those positions in two or

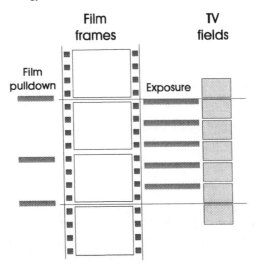

Figure 7.4 *3:2 pulldown for conversion of 24 f/s film on 30 f/s video systems. (Reproduced with permission from [3].)*

three fields, the reproduced motion becomes uneven. An ideal telecine system should correct this problem. It can be done with digital technology.

7.3.1.2 Digital Frame Rate Conversion

With digital video technology, it is easy and inexpensive to store the video data for one or

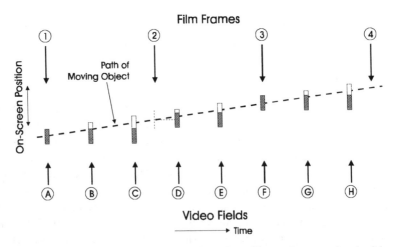

Figure 7.5 *Effect of 3:2 processing on a moving object. Dotted boxes show actual path of the moving object; shaded boxes show where they appear with 3:2 processing.*

more video frames in a RAM memory. Data is written into memory at the film frame rate and read out at the video frame (or field) rate. The result is electronic frame rate conversion.

A simple algorithm for digital frame-rate conversion is to use the 3:2 approach already described. Even if that is all that is done, it is advantageous to do it digitally because it disconnects the film pulldown from the video system. As Figure 7.4 showed, the pulldown of film frames must occur at precise points to keep the film frame stable for the two or three exposures to the video camera. This requires an intermittent film mechanism (see Section 7.4.1) that is expensive, unreliable, and potentially damaging to the film. With a digital memory, film pulldown can occur at any time or the film can even move continuously using a line-scan pickup device or an area-scan pickup with an appropriate fast-exposure system such as a strobe light. The multiple frames in the video system are achieved by reading the memory twice or three times before updating it from the film.

However, 3:2 is restricted to the exact 24:30 frame rate relationship, which may not always be the desired speeds. Having the digital memory, a much better system can be built by interpolating frames, which removes any restriction between the input and output frame rates. This technology has been highly developed for video standards conversion and frame synchronizers. Two or more adjacent film frames are stored in memory and frames at in-between times, as required by the output frame rate, are calculated by combining corresponding pixels from the surrounding frames.

That is not as easy as it sounds because the approach breaks down on moving objects. For pixel interpolation to work on a moving object, the object must be located in each frame and corresponding pixels *on the object* should be processed instead of corresponding pixels based on screen location alone. This is called *motion compensation* and it gets very complicated. The basic idea is to break the picture up into blocks and search an area in each new frame to determine if any of the blocks from the previous frame can be found at a different location in the new frame. If so, a motion vector is associated with the block in the new frame to specify where that block can be found in the previous frame. The interpolation processing then takes the motion vectors into account when computing the interpolated frames. A full description is beyond the scope of this book, but it is covered further is Section 4.10.4.1, and complete information can be found in the literature of the video compression field [4].

Frame interpolation with motion compensation is relatively new in telecine systems. However, it is undergoing continued development and cost reduction and it should be more common in the future.

7.4 FILM TRANSPORTS

Playing a film involves unreeling the film to present it frame-by-frame to a playing station (*gate*) and rolling the film up again afterward. At the end of the film, it must be rewound back onto its original reel. These seemingly simple tasks define the purpose of a film transport mechanism. There are many other considerations:

- Film speed must be smooth and precisely controlled—film is perforated for speed control; there is always an integral relationship between the perforations and the frames, e.g., four perfs per frame. Some transports control film speed at critical points by directly driving a sprocket that is engaged with the perforations. However, even in a continuous-motion transport where the film may be driven by a capstan, perforations must be sensed to control speed because of film shrinkage or stretch. For telecine applications, the film pulldown rate usually must bear a fixed relationship to the vertical-scan rate of the video system. This implies servo control of the driving elements of the film transport.

- Variable-speed operation—this is needed for film loading, shuttle, and rewinding. High-speed operation should not cause film instability or damage.

- Multiple film sizes and formats must be accommodated with easy and fast change-over between modes. Most telecine manufacturers offer interchangeable gates, rollers, sprockets, and other components to adapt their transports to 16- or 35-mm film in various formats. An important consideration is how easy or difficult it is to make the changes for different film sizes. Various automatic setup and alignment features are often provided.

- The operation of the reading gate must present a stable image to the imager. In intermittent transports, the method of registering the film to an exact position in the gate is important. In all transports, film must be precisely guided through the reading gate to avoid side-to-side film instability (called *weave*) and it must be position at the exact focus point in the optical system.

- Film must be unwound and wound from/to reels with controlled tension. Most transports have servo control of the reel tension.

- Extremely smooth and stable motion must be established at the point of reading sound tracks. Since film motion provides the time base for the film sound, non-uniform motion will cause wow and flutter of the sound. Sound heads often run the film around a flywheel to achieve the necessary stability.

- Rewind operation must be accurate and fast. This is another purpose for the reel tension system and the film guidance system in the transport. It may be necessary in some designs to release the film from some guidance elements for rewinding.

- Film must be protected from damage to edges or emulsion. The emulsion surface of active picture areas of the film should never be allowed to slide across any guiding elements of the transport. Rollers may be used, but they should be driven by the perforation areas of the film and they should not be allowed to slide on the film while accelerating or decelerating. One approach is to design the roller to only contact the film at the perforation area by slightly reducing the diameter in the frame area of the film. Also, to minimize sliding of the roller during acceleration, its rotating mass can be reduced by having only a thin cylinder rotate, instead of the entire roller. Edge guiding should exert minimal force on the film for guidance. This means that transport elements should be precisely aligned for perpendicularity or parallelism so that the film will naturally flow through the transport almost without edge guiding.

- Loading and unloading of the film transport must be convenient and easy. On loading, film reels are lifted onto hubs, which should be at a convenient height. Film has to be threaded through the film path; this should be straightforward and the film path elements should move out of the way where necessary for easy threading.

- The mechanism must be mechanically stable and easy to maintain. Manufacturing to the precision called for by the above-mentioned requirements calls for well-stabilized castings and components and careful alignment. The structure and components should be strong and well fastened so there will be no movement under normal use conditions. Similarly, the transport should be environmentally stable so it will stay in alignment during normal temperature or humidity changes. Transports also tend to get dirty because of debris on the film or in the environment; they should provide easy access for cleaning of all critical areas.

Film transports differ significantly between those that have intermittent film motion at the gate and those that have continuous film motion.

7.4.1 Intermittent Transports

The purpose of an intermittent film transport is to move the film in the reading gate one frame at a time, as fast as possible, while providing a stable frame for reading during a high percentage of the frame time. This type of transport is used in telecines that have pickup methods requiring a stable image for a significant portion of the frame period. That includes area-scan CCDs and photoconductive vacuum-tube imagers. Figure 7.6 is a layout diagram for an intermittent-motion film transport.

7.4.1.1 Pulldown

The process of moving the film in an intermittent mechanism is called the *pulldown*; it is usually done with a *claw mechanism* having two or three teeth that intermittently engage the film's perforations for pulldown, perform the pulldown, and reference the film to an exact position in the gate. Once this has completed, the film frame in the gate is available for exposure until the next pulldown cycle begins.

During the pulldown, the film should be free to move rapidly through the gate area under control of the claw. However, elsewhere in the transport, the film must move smoothly for reeling and sound pickup. This is accomplished by having a loose loop of film on either side of the gate so the claw can pull film from and to the loops without disturbing the film motion anywhere else in the path. If, for any reason, there is a bad pulldown, such as may be caused by damaged perforations, the loops of film may become "lost," meaning that the film becomes tight instead of loose in the loop area. This will cause continued failure of the pulldown until corrected. Film transports must have means for "loop restoring" to automatically recover from such a fault.

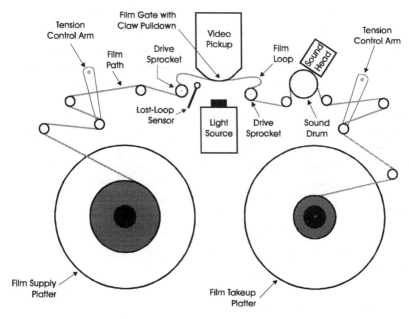

Figure 7.6 *Layout of a typical intermittent-motion film transport.*

Intermittent mechanisms are used in theater film projectors; they have been highly developed for that application. To the extent that telecine applications do not require faster-than-normal pulldown speeds, the theater experience applies. However, at best, an intermittent mechanism is tricky and the trend in telecine is to avoid them in favor of continuous-motion film transports.

7.4.2 Continuous-Motion Transports

In a continuous-motion film transport, the film moves smoothly through the transport and the gate without any intermittent motion. Whereas, in intermittent transports, the film is usually sprocket-driven by the perforations, in continuous-motion transports, the film is usually capstan-driven. The capstan is a precisely controlled wheel that contacts the base side of the film to drive it at constant speed. Figure 7.7 is a layout diagram of a continuous-motion film transport.

However, the perforations must still be used for positioning the film at the gate; this requires a servo controller that senses the rate of flow of perforations and controls the capstan speed to achieve the correct film speed. Perforations may be sensed by a sprocketed idler that is driven by the film; a tachometer wheel coupled to the sprocketed idler and optically sensed can deliver an output whose frequency is related to the film speed. This works well with good film but it is subject to the problems of handling film whose perforations may be damaged. An alternative approach is to sense the perforations directly with an optical sensor looking at the film. This has the advantage that it does not depend

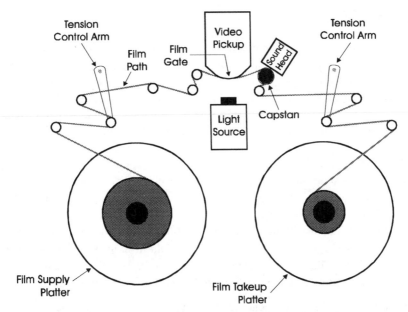

Figure 7.7 Layout of a typical continuous-motion film transport.

on the mechanical engagement of perforations with a sprocket and electronic processing can deal with the problems of damaged perforations. However, there are pros and cons of these approaches and it is difficult to make an absolute conclusion about which is better. Either approach, when well designed, will operate successfully in 99% of the cases.

A continuous-motion transport has a clear advantage of being easier on the film. Intermittent mechanisms that must accelerate and decelerate the film via the perforations, inherently have more possibility of damaging the film in use. For the best treatment of precious master films or old films, continuous-motion is the clear choice.

7.4.2.1 The Film Gate

In all film transports or projectors, the film must be precisely positioned at the gate to ensure proper focus and image lateral stability. For control of focus, the film is usually made to slide through the gate in close contact with a supporting surface against the emulsion side of the perforation area of the film. This controls the position of the image on the surface of the film independently of film thickness. There must be enough tension on the film at this point that the film will remain flat in the gate. Some designs use a curved gate, where the supporting surfaces of the gate follow a curve along the film path. This helps to establish uniformity of film position across the gate, even with somewhat damaged film. However, it requires special optical considerations to maintain focus all over the cylindrically curved surface. Note that line-scan pickup devices that are viewing the film in a single line position can use a curved gate with no special optical considerations.

7.4.3 Optical Considerations

A film-pickup optical system has numerous considerations, which are listed here:

- Illumination—the optical system must ensure that the entire active area of the film gate is uniformly illuminated, both for intensity and color. Nonuniform illumination will cause shading, which might be corrected electronically, but that uses up some of the system's dynamic range. The optical system should provide the best possible uniformity; then electronic shading correction need only correct for what remains.

- Heating of the film—because most film imagers have low sensitivity, a large amount of light must be focused on the film gate. This can excessively heat of the film if the film stays too long in the gate, causing physical changes in the film or even damage in extreme cases. Features such as heat-shielding filters in the illumination path and reducing illumination when the film stops in the gate are essential.

- Focus—as already explained, the entire area of the film gate must be in focus. Special design of film optical systems can provide good depth of field to achieve focus over the entire area and to maintain that with different films and over time.

Telecine optical systems vary depending on the type of sensing used. Therefore, specific optical designs are covered in the following sections on types of cameras.

7.5 FLYING-SPOT CAMERAS

Telecine cameras have long used flying-spot pickup because it has advantages in the lack of color-registration problems and it is simple to operate. Of course, flying-spot is a non-storage pickup method that has low sensitivity, but the controlled optical environment of film pickup provides high-enough optical efficiency that the flying-spot is practical. Flying-spot also has the advantage that it can use a continuous-motion film transport, even without any digital memory. However, it can work even better with digital memory, which is no longer much of a cost consideration with today's memory technology.

Figure 7.8 shows a block diagram and optical layout of a typical flying-spot telecine. A monochrome cathode-ray tube (CRT) has a raster scan that illuminates the film in the gate with white light. The light passing through the film is collected and passes through an RGB light-splitter to three photomultipler tubes, which deliver R, G, and B video signals to the electronics.

7.5.1 CRT and Optical System

In addition to the need for high intensity, a flying-spot CRT must have fast persistence, which is the decay of light from an area of the phosphor after the beam has moved on. Persistence of the CRT causes smearing of the video, which has to be corrected in the signal electronics. Even the best CRTs require persistence correction.

The color of the CRT should have a broad spectrum, as close to white as possible, although that also can be corrected electronically. The intensity requirement overrides the color requirement and phosphor design favors efficiency and light output. Another phosphor consideration is graininess; phosphor graininess or other nonuniformities of efficiency will cause fixed-pattern noise in the video. A still further phosphor consideration is burning; high-intensity phosphors tend to change their characteristics in the area scanned by the raster. This becomes visible if the raster moves slightly on the face of the CRT. Some flying-spot telecines deliberately move the raster slowly to prevent one area from burning too much. Of course, moving the raster causes moving of the scanned area of the film in the gate and this has to be compensated electronically or optically. Even so, the end-of-life of a flying-spot CRT is usually caused by burning.

The light emitted from a CRT phosphor comes out in all directions, including back into the tube envelope. Optical efficiency is improved by designing a mirror structure into the phosphor to reflect light going inside the tube outward instead. Another optical feature is the *fiber-optic faceplate*, which causes the phosphor light to come out in a nearly parallel beam instead of scattering in all directions in front of the tube. These features can substantially increase the light output from a flying-spot CRT.

The optics on the illumination side has the objective of collecting as much light as possible from the raster on the face of the CRT and focusing it at the correct size onto the film in the gate. On the other side of the film, the optical objective is to collect as much of the light coming through the film as possible and focusing that onto the photomultiplier tubes in the light splitter. Both optics should be designed to deliver uniform illumination or light collection over the area of the film. Any nonuniformity of the light collection shows up as shading in the reproduced video. With positive film, it is white shading; with negative film, it will be black shading. Either must be corrected in the electronics for best reproduction.

7.5.2 Scanning Methods

Before the days of digital memories, flying-spot scanners used continuous film motion with a scanning method called jump-scan that tracked the film motion and produced the 3:2 processing at the same time. This worked, but it was tricky and difficult to maintain. Because jump-scan required the raster area to change dynamically on the CRT phosphor, the phosphor surface was not uniformly illuminated on the average. This resulted in nonuniform burning of the phosphor, which had to be corrected electronically as it developed during the CRT life.

Modern flying-spot telecines still use continuous film motion, but the frames are read into digital memories where 3:2 or other frame-rate processing, such as interlacing, occurs. Theoretically, this could be accomplished by just having a single-line scan on the CRT, but that would result in rapid burning of the phosphor and is unacceptable. The CRT must be scanned with some sort of raster to prevent rapid burning. That is accomplished by speeding up the horizontal scanning of the CRT so that the total lines of a video frame (525, 625, and so on) are scanned in less time than the film frame time (1/24 s). The result is that the film motion does not provide all the vertical scanning; the rest is

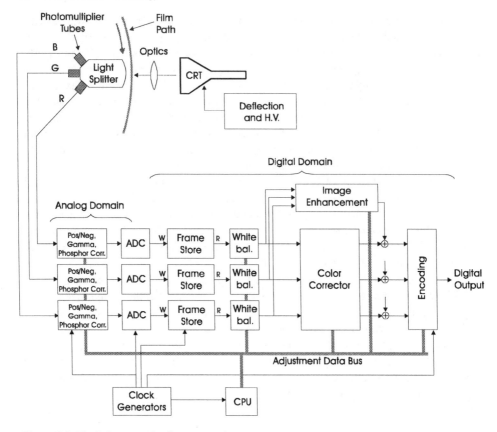

Figure 7.8 *Block diagram of a flying-spot telecine system.*

made up by a small vertical scan on the CRT. Usually a speedup ratio of 4:3 is used, meaning that the film motion provides ¾ of the vertical scanning and the CRT provides the other ¼. The conversion back to the correct horizontal scan rate occurs in the readout of the digital memory.

7.5.3 Flying-Spot Processing Electronics

As shown in Figure 7.8, in addition to the usual processing of video in any RGB camera that was described in Chapter 4, a flying-spot telecine requires the CRT phosphor correction, positive and negative film processing, and the digital frame-storage units. These features as well as the color correction and gamma correction require different considerations for telecine, which are discussed in Section 7.7.

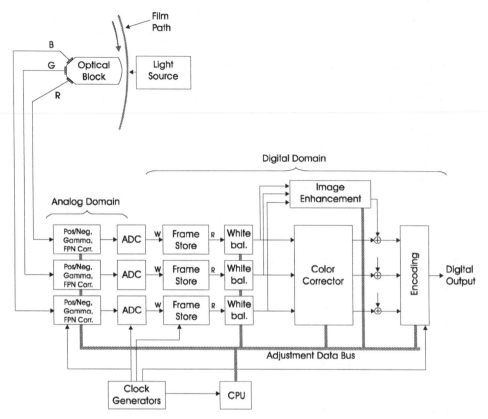

Figure 7.9 *Block diagran of a CCD line-scan telecine system.*

7.6 CCD TELECINE CAMERAS

The same area-scan CCDs used in live-pickup video cameras may be used in telecine systems. However, because the film motion can easily provide vertical scanning, there is also the possibility of using a single-line scan CCD. This is simpler and may be less expensive, but it introduces its own set of considerations. Both types of CCD telecine cameras are discussed in this section.

7.6.1 CCD Line-Scan Telecine Cameras

Because of their relative simplicity, line-scan CCD telecines have been more popular than area-scan CCDs. Figure 7.9 shows a block diagram for a CCD line-scan system.

The film is illuminated in the gate by an incandescent light source, whose optics strive for uniform illumination along the line in the gate where the CCDs will be looking. On the other side of the film, the illuminated line in the film gate is focused onto three CCDs through light-splitting optics, with the film frame width of the image on the CCD surfaces matching the CCD's active sensing width.

The line-scan CCDs typically have 1,000 to 1,500 pixels per line, which yields sufficient resolution for all standard video systems and almost enough for HDTV. Normally, the CCDs are clocked at a rate that yields the desired number of scan lines in the time of passage of a film frame through the gate. Some CCD telecines provide faster-clocking modes that will scan only part of the film frame as it goes by; this gives a magnified image of the film frame that still has full video resolution. To the extent that the film has more resolution than the video system, as large-format films do, considerable magnification is possible before the system MTF begins to deteriorate.

The rest of the signal processing is similar to a flying-spot camera except that no phosphor correction is required. Signal processing considerations are discussed further in Section 7.7.

7.6.2 CCD Area-Scan Telecine Cameras

It might seem that a line-scan CCD would be the ultimate approach for a film camera, but there are certain advantages to area-scan devices that become important in certain applications. These are:

- Sensitivity—one would think that light sensitivity would not be an important consideration in a film camera, but some films can be very dense, especially an overexposed negative. This can cause as much as 100:1 loss going through the film and one would like to increase the illumination to still get a good signal from the camera. However, there are limits to illumination: in a flying-spot telecine, there is only so much light the CRT can deliver; in other telecines, the illumination is limited by power and heating of the film. Often the camera sensitivity must be cranked up to get a good picture from dense films. Because of the much longer integration time in area-scan CCDs compared to line-scan CCDs, the former have much higher sensitivity and light or optical efficiency would never be a problem. Of course, exploiting the higher sensitivity of area-scan devices requires intermittent film motion. If a continuous-motion transport and strobe illumination is used with area-scan devices, area-scan sensitivity is the same as a line-scan camera.

- Film speed uniformity—systems that scan the film by means of film motion depend on uniformity of that motion during a scan to maintain picture geometry. This is not a problem in systems that scan each frame while the film is stationary, or with systems that capture the frame with a strobe light. Both of these require area imagers. With good film, motion uniformity is not a problem and all types of telecine systems give good pictures but, with old or damaged film, an area-sensor telecine may give better results in terms of picture stability.

- Availability—area-imaging devices and assemblies of many types are made in large quantities for live-pickup cameras. It is reasonable to consider adapting one of these sensing assemblies to a telecine system. The advantages of volume production, low cost, reliability, and performance developed for live cameras can thus accrue to telecine cameras, which is a small market that could never support such developments itself.

7.7 SIGNAL PROCESSING FOR TELECINE

Much of the signal processing in a telecine is the same as live-pickup video cameras, which was discussed in Chapter 4. However, some differences have alread been noted and they are discussed further in this section.

7.7.1 Negative Film Processing

As noted above, negative film poses many more signal processing problems than simply a reversal of signal polarity. Because white and black are reversed in the negative areas of the signal chain, the same setup automation used for positive signals does not work correctly. For example, it is common in positive-signal chains to provide an optical black reference in the imager that can be sensed in the signal chain to establish an absolute black level. In practice, the scene black level is generally very close to the reference level and the reference is sufficient to black balance the camera. In a negative film processing chain, the same black reference in the imager results in a white reference level in the signal that is far different from scene white because of the minimum density of the film negative. It is not suitable for level setting or for balancing; the latter is because the minimum density of film negatives is deliberately colored to provide a degree of masking to an eventual color-printing process. This must be taken into account in the color balancing of a negative-film processing chain.

7.7.2 Gamma

Because positive film has a high gamma and negative film has a low gamma (which becomes high when the polarity is reversed), film pickup requires much heavier (lower number) gamma correction than live pickup cameras. Whereas live cameras usually use about 0.45 gamma, film cameras may need gammas as low as 0.25 to 0.3. This results in much higher gain expansion in the black regions and a corresponding increase in noise. Because of this, film cameras need even better SNR from the imaging process than a live camera for equivalent overall noise performance.

With the extreme black stretch applied by low-gamma processing, digital conversion done before gamma correction must have more bits per sample than a live camera. Because of that, many so-called "digital" telecine systems still do gamma correction in the analog domain. Even so, 10- to 12-bit conversion is common.

The highlight handling capability of the electronics also differs in a film camera. With positive film, the brightest elements of the image are caused by the minimum density in the film image. However, a greater highlight could be produced by a hole in the film or a severe scratch in the emulsion, where the total illumination of the light source can come through. Because minimum densities of positive films are usually 0.5 or less, highlights seldom will go above 3:1 or so, unless the film is extremely dense. This is easily handled by conventional means.

Negative film reproduces highlights with high film density. A scratch on the film will produce a black line in the video. Dirt on the film will show up as white. Even though highlights are not a large problem, dirt and scratches are generally more visible in the video when reproducing negative film.

7.7.3 Color Correction

Reproduction of film on video is subject to considerable variation of color. This may be due to film aging problems or it may simply be due to film variations that are not objectionable when the film is projected but become more serious when the film is transferred to video. These problems can vary from scene-to-scene on the film or even sometimes from frame-to-frame. High-quality film-to-video transfers generally require the use of a *color corrector* system that is capable of performing color manipulation on a frame-by-frame programmed basis. Such capabilities are generally not built into telecine systems; rather, telecines provide an interface for a third-party manufactured color corrector.

Film color correction is a sophisticated business and requires a skilled expert for the best results. A person who has learned this skill and practices it professionally is known as a *colorist*.

7.8 SOUND PICKUP CONSIDERATIONS

Film sound may be provided by tracks running along the side of the film between the frames and the perforations. These sound-on-film records may be either optical or magnetic. There are some limitations to sound-on-film performance because the space available on the film is small and the film motion stability may not be the best. For highest quality, film systems use *sepmag*, which is a separate magnetic sound recorder that operates synchronized with the film. Telecine systems are called upon to support all these varieties of sound recording.

7.9 FILM SCANNERS

There is another class of telecine system that is not necessarily designed for real-time film transfer. These systems, called *film scanners*, capture a high resolution image of each film frame, with resolutions of 2,048 × 2,048 or higher, and transfer them to separate digital files for each frame for storage on digital hard disks or other media. These systems are used for archiving and for digital processing of film for special effects. For example, scenes of feature movies that undergo digital effects processing must be transferred to digital by a film scanner. The output of the scanner is stored in the digital effects system.

When the effects are completed, the results are transferred back to film for editing into the rest of the movie. Obviously, this requires the highest possible performance in the film scanner.

REFERENCES

1. *SMPTE Journal*, Vol. 105, No. 9, September 1996. This entire issue is devoted to the Fox Movietone News digital archiving project.

2. SMPTE standards are available from the Society of Motion Picture and Television Engineers, 595 W. Hartsdale, Ave., White Plains, NY 10607-1824, or http://www.smpte.org.

3. Luther, A. C., *Principles of Digital Audio and Video*, Artech House, Norwood, MA, 1997.

4. Ibid, Chapter 9.

8

Camcorders

In the early days of video (television), it was not possible to go out in the field and shoot video to bring back for later use in the studio. When that was needed, the only solution was to use motion picture film. With the development of magnetic video recording, it became technically possible to take cameras and recorders into the field and accomplish the same thing that could be done with film. However, early video cameras and recorders were large, heavy, complex, and power-hungry. Taking them in the field was a major operation. With the steady advance of electronic technology that made things smaller, lighter, higher-performing, and less expensive, by the 1970s video cameras and recorders became small enough that it was conceivable for one person to carry them both into the field. At that point it became obvious to combine a camera and recorder into one unit—a *camcorder*. Now, video could be shot just as conveniently as film, and the other advantages of video made camcorder shooting highly competitive with film shooting for any video applications that required immediacy, such as television news.

Camcorders have come a long way from the backbreaking devices they were at the beginning. This chapter is about camcorders and their technology.

8.1 TYPES AND USES OF CAMCORDERS

Camcorders range from the small units built for home use and costing only a few hundred dollars to the larger and more sophisticated units used for professional news or production shooting. These latter camcorders can cost tens of thousands of dollars and may include nearly all of the features of studio production cameras. Digital camcorders are now appearing in all markets and promise to take over much of the market in the next few years. Table 8.1 lists three classes of camcorders and gives some of their features and specifications.

The table may be somewhat confusing because it embraces a wide range of cameras, even within each class. Further, it can be seen that many features contained in home

Table 8.1
Camcorder Characteristics

Feature/Specification	Home	Semi-Professional	Professional
		Market	
Imaging technology	1-CCD	1-CCD, 3-CCD	3-CCD
Imager size (diagonal-in)	0.25, 0.33	0.33, 0.5	0.5, 0.67
Imager pixels	250,000 – 350,000	350,000 – 500,000	400,000 up
Recording technology	VHS, 8 mm	S-VHS, Hi-8, DVC	Betacam, DVCPRO
Recording time (min)	30 – 360	30 – 360	30 – 60
Configuration	Hand-held, shoulder	Hand-held, shoulder	Shoulder
Weight range (lb)	1 – 2	2 – 10	10 – 15
Zoom range (optical)	8× to 12×	10× to 15×	10× to 20×
Zoom range (digital)	20× to 100×	20×	None
Viewfinder	Mono CRT, color LCD	Mono CRT, color LCD	Mono CRT
H resolution (camera), TVL	300 – 400	350 – 600	600 – 900
H resolution (VCR), TVL	250 – 300	300 – 400	500 – 600
Minimum illumination (lux)	1	1	2 to 4
SNR (camera), dB	40 – 45	45 – 55	55 – 64
Special effects editing	Yes	Yes	No
Image stabilization	Yes, at high end	Optional	No
Docking	No	No	Optional
Triax capability	No	No	Optional
Remote control	VCR only	VCR only	Full—optional
Price range ($)	500 – 1,500	1,500 – 5,000	over 10,000

cameras are not present in the much higher priced professional cameras. This apparent anomaly results from the different attitudes of the markets about trading off signal performance for operating features (e.g., digital zoom is generally not provided in professional camcorders because it trades off resolution performance). On the other hand, price is an overriding requirement for the home market and performance trade-offs are made in that interest.

Another consideration about camera features is that a home camcorder operates by itself and is not part of a system. Therefore, it may contain features, such as editing or special effects, that normally are provided by the rest of the system in a professional environment.

Although "home" digital camcorders have come on the market recently, they are priced in the semi-professional range, and therefore are included in that column of Table 8.1. It is expected that prices of digital camcorders will be coming down and some of them soon may be in the home price range. Other factors influencing the behavior shown in this table will be discussed in the rest of this chapter.

8.2 VIDEOTAPE RECORDING

Recording technology is a major consideration to achieve the overall physical and electrical objectives of a camcorder. This chapter does not attempt to give complete coverage of recording technology, that is covered in the literature [1,2], but it will be introduced here just enough to support the discussion in the rest of this chapter.

8.2.1 Tape Recording Fundamentals

All camcorders use magnetic recording, generally on tape, although the use of magnetic disks is emerging in certain applications (see Section 8.3). Magnetic tape is a flexible plastic substrate on which a magnetically active coating is placed. Tapes have widths in the range from 0.25 to 0.5 in and a thickness from 0.0005 to 0.001 in. Tape is rolled onto reels in a cassette that makes the tape easy to handle and protects it from rough handling and the environment. For recording or replay, the tape cassette is inserted into a *tape transport* (sometimes called a *tape deck*). In operation, the tape transport extracts a loop of tape from the cassette and threads it into a tape path containing the elements that record or replay audio and video. Additional tape path elements are required for the purpose of controlling the handling of the tape in the transport.

Recording is accomplished by locally magnetizing the tape's magnetic coating in a pattern of *tracks* that generally are in the form of diagonal lines across the tape as shown in Figure 8.1(a). This is known as a *helical-scan* recording and it is made in a tape transport that has a scanner assembly including a rotating wheel that is set at an angle to the tape path as shown in Figure 8.1(b). The tape is supported in a path that wraps around the wheel and *magnetic heads* on the rotating wheel record or replay the tracks on the tape as the tape moves past the wheel. Helical-scan recording allows a much higher *head-to-tape scanning speed* (also called the *writing speed*) than the speed of the tape alone; this high speed is essential to achieve the bandwidth needed for video recording.

Helical-scan recorders may be classified according to the angle of tape wrap around the scanner drum. Some scanners try to wrap the tape almost entirely around the drum, leaving only a narrow angle where the tape enters and leaves the drum as shown in Figure 8.1(c). This is known as *omega-wrap* and, with this approach, a video recorder needs only a single head on the scanner. In analog systems, the short gap in recording where the head passes from the end of one track to the start of the next track can be synchronized to occur in the vertical blanking interval of a video signal. However, keeping the time gap short enough to stay in the VBI makes single-head omega-wrap mechanisms complex and tricky and they are generally not used in camcorders.

An alternative is to open up the time gap to nearly half of the time and place two heads on the scanner drum as shown in Figure 8.1(d). The heads can be alternately switched so that the signal is always derived from the one of them that is currently scanning the tape. Thus, there is no gap in recording. The mechanism is also simpler, but it has the disadvantage for camcorders that, for a given writing speed, a two-head scanner must be nearly twice the diameter of a one-head scanner. That makes the entire tape transport larger.

The magnetic tape medium is highly nonlinear and some form of modulation is required to record video or audio. In analog systems, frequency modulation is generally used for video and a special technique known as *bias recording* is used for baseband audio. Analog audio recording requires a low enough bandwidth that audio tracks may be placed along the edges of the tape. The audio writing speed thus is equal to the tape speed.

In digital systems, there are many techniques of modulation available for recording of bit streams [4]. Generally, digital recorders multiplex audio and video into the same tracks, since multiplexing is so easily done with bit streams.

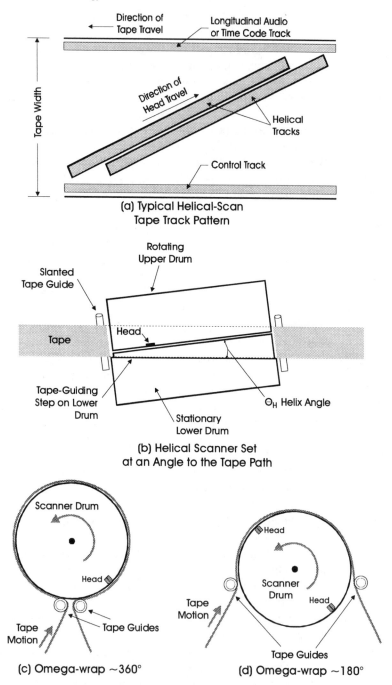

Figure 8.1 Helical-scan recording. (Adapted with permission from [3].)

8.2.2 Videotape Recording Formats

Video recording formats require a supporting infrastructure, including availability of cassettes and editing recorders. Even in the home market, the consumer would like to be able to play tapes from the camcorder on his or her home VCR. Thus, camcorder formats are compatible with home VCRs or with those professional studio recording formats that are capable of being built small enough for camcorder use. Note that popular studio recorders, such as SMPTE Type C or the SMPTE D-series, are too large for camcorder use. Table 8.2 lists the most common camcorder recording formats and some of their characteristics. Because home-type VCRs generally use smaller formats, they have been the basis for recording in camcorders, even the professional ones. Basic mechanical components of home recorders have been adapted to professional camcorders, but the modulation parameters are modified to achieve higher picture quality.

The following paragraphs explain the items in the table.

- Tape width—this has been trending down over the years as *recording density* has grown and more information can be recorded per unit area of tape. Current widths are 0.5, 0.31 (8 mm), and 0.25 in.
- Tape speed—this is going the same direction as tape width. Lower speed means less tape used per minute and longer recording time for a given cassette size.
- Recording time—this is likely to remain in the range of a few hours. Formats designed for portable use will sometimes trade off recording time for cassette size, as in the VHS-C format, which has the same tape track size and recording pattern as VHS but with a smaller cassette. In the long run, however, all systems strive for 1 hour or more recording time. Note that the VHS format provides a choice of tape speeds; this is a trade-off between recording time and picture quality.
- Helical-scan parameters—the drum diameter and the drum rotational speed determine the video writing speed. The writing speed and the video modulation method determines the bandwidth. As basic recording technology improves, writing speeds go down, less tape area is used per unit time, and tape transports become smaller and lighter. However, changes in these fundamental parameters are not made lightly because they require new transport designs and possibly new cassette or tape designs as well. Such changes upset the infrastructure and introduce problems of compatibility with older equipment.
- Drum rotation speed—with analog formats, drum rotation must be synchronized with the video vertical scan so that head switching occurs during VBI. That fixes the drum diameter for a given writing speed. Digital recorders, however, have no restriction on the location of head switching and drum diameter can be reduced and rotation speed increased for a given writing speed. This trend is clear in the numbers for DVC, which is a digital format that has been designed from scratch.
- Video modulation—in analog recording, the *color-under* format has predominated in small recorders. This method uses FM modulation for the luminance signal and records the chrominance components in a narrow frequency band be-

Table 8.2
Camcorder Recording Formats

Item	Analog Formats				Digital Formats		
	VHS	8 mm	SVHS	Hi-8	Digital Betacam	DVC	DVCPRO
Tape width, in	0.5	0.31	0.5	0.31	0.5	0.25	0.25
Tape speed, in/s	1.33/0.66/0.43	0.56/0.28	1.33/0.66/0.43	0.56/0.28	3.8	0.75	1.33
Recording time, min	120/240/360*	120/240	120/240/360*	120/240	40	120/240	63/120
Helical tracks							
Drum diameter, in	2.44/1.63	1.57	2.44/1.63	1.57	2.93	0.855	0.855
Rotation speed, rps	30/45	30	30/45	30	74.925	150	150
Number of heads/track	2/4	2	2/4	2	2	2	2
Total heads on drum	2/4	2	2/4	2	2	4	4
Writing speed, in/s	229	148	229	148	689	201	201
Video modulation	Color-under	Color-under	Color-under	Color-under	4:2:2×10 bits	4:1:1 compr.	4:1:1 con
Y bandwidth MHz	3	3	4	4	5.75		
Color bandwidth MHz	0.4	0.4	0.4	0.4	2.75		
Audio tracks							
Number of tracks	2 PCM	2-stereo	2 PCM	2-stereo	2	2 PCM	2 PCM
Bandwidth					20 – 20,000 Hz		

* Full-size cassette. A small cassette used in some camcorders is called VHS-C and SVHS-C. Playing times are 20/40/60 for these.

low the FM spectrum. This inherently has a narrow chrominance bandwidth. In a larger format with a higher writing speed, the FM bandwidth can be made great enough to record luminance and chrominance together. The modern solution is to go to digital recording, where a small mechanism can still have a high writing speed, and the digital modulation can deliver high picture quality.

- Audio—video recorders always have audio recording capability, generally at least two channels for stereo. In early VCRs, audio was recorded by conventional bias recording technology on linear tracks along the edges of the tape. However, as video recording advances allowed tape speeds to become slower, the quality of linear-track audio became too poor. Digital audio was then added to the video tracks in all formats (both analog and digital) so that audio quality could remain competitive with other audio systems.

The recorder considerations discussed above have a significant effect on both physical characteristics and signal performance of camcorders.

8.3 HARD DISK CAMCORDERS

As digital video technology has advanced, it has become common to record video on computer hard disk drives. This capability has made possible the use of computers for *nonlinear editing* of video, which has revolutionized the video postproduction field. With digital video compression, even the hard drives in small personal computers can record and replay video. The numbers for video data rates and hard drive storage capacity are beginning to support recording of tens of minutes of video on small, portable hard drives that could be used in a small camcorder. That promises soon to be an important class of camcorder.

8.3.1 Hard Disk Fundamentals

Figure 8.2 shows a hard disk drive. One or more rigid platters (disks) rotate at high speed (60 rps or more) and a magnetic head is mounted on a rotating arm so it can move approximately along a radius of the platter. The head is a special design that does not actually contact the platter surface; it *flies* a very small distance above the platter on a film of air produced by the platter rotation. Digital data is recorded on concentric tracks on the platter and the head arm is driven by a stepping motor that allows it to move rapidly from track to track. Under computer control, the head can be rapidly moved (this is called *seeking*) to write or read any track on the platter.

In personal computers, hard drives are managed by a *file system*, which is software that allows blocks of data (files) on the disk to be addressed by giving them names (file names.) Files may be of any size and there may be thousands of them on a large hard disk. A file on the disk is accessed physically by seeking to a track and a position within the track (called a sector), but the file system hides all this from the user by placing a *directory* file on the disk that associates names with tracks and sectors. Often, there are multiple directories, which also may be hierarchical (directories within directories.) Since a

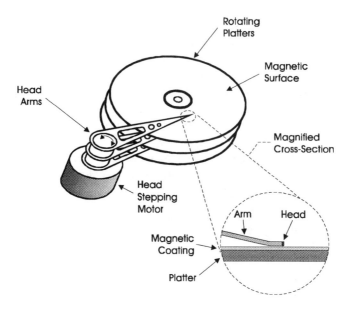

Figure 8.2 *Magnetic hard disk drive.*

magnetic hard disk is a rewriteable medium, files that are no longer needed can be erased and the disk space reused for new recording.

Hard disk storage capacity is a rapidly moving target. At this writing, desktop PC hard drive capacities are in the range of 4 GB per drive and trending upward. PC hard drives are of such size and weight that they could be used in a shoulder-held camcorder if suitably ruggedized. However, even with these capacities, an adequate recording time would require some video compression, which is not desirable for professional use.

Portable drives, such are used in laptop personal computers have much less capacity—200 to 300 MB. These latter drives are the ones most adaptable to hand-held camcorders. At home video quality levels, video can be compressed to about 10 MB per minute, so a 200 MB hard drive can store 20 minutes of such video. That is a minimal amount, but hard drive technology is moving rapidly and portable-drive capacities soon will be over 1 GB, which will give enough recording time for almost any application.

A plug-in card standard for portable computers was developed a few years ago, originally for memory expansion, but now adapted to all types of peripherals, including hard disks. It is called PCMCIA, for *Personal Computer Memory Card Industry Association*, which is the industry body responsible for the standard. PCMCIA hard disks have capacities as large as 350 MB at this writing.

Recordings on hard disks are made on platters that are not removable from the drive. Unless the entire drive is removable, as in PCMCIA drives, the data can only be retrieved from a hard disk camcorder by communicating it to another storage device, such as the hard drives in a PC. Thus, data communication is an important consideration for hard disk camcorders.

Figure 8.3 *The Hitachi MPEG Camera.*

8.3.2 A Hard Disk Camcorder

The first commercially-available hard disk camcorder is the Hitachi MPEG Camera, shown in Figure 8.3. This hand-held unit uses a 260 MB PCMCIA removable hard disk and can store up to 20 min of video and audio using MPEG-1 compression.

The key to the practicality of a hard disk camcorder is video compression. The MPEG Camera uses MPEG-1 compression (see Section 4.10.4), and compresses video to about 10 to 13 MB/min. This delivers home-video picture quality at a resolution of 352 × 240 pixels, which is acceptable for home use and many applications on PCs, the Internet, or the World Wide Web. Hitachi developed a single IC for this camera that performs the MPEG compression and decompression.

Higher video quality for professional applications would require the use of MPEG-2 compression and data rates of 50 to 100 MB/min. That is not yet practical for a hard disk camcorder, but it will surely appear as soon as portable hard disk capacities grow into the 1 GB range.

The MPEG Camera CCD is a ¼-in 390,000-pixel device with a resolution of 704 × 480 pixels. The full resolution is available in still-picture mode. Using this with JPEG compression, up to 3,000 still images can be stored on the hard disk. Other options for the disk files include still pictures with audio or audio alone (up to 4 hours). Any combination of video, stills, or the other file formats can be combined within the 260 MB capacity of the disk.

8.3.2.1 MPEG Camera Features

The MPEG Camera demonstrates some of the unique features that become possible with a hard disk camcorder. In many respects, they are similar to the features of digital still-image cameras (see Chapter 11).

- The MPEG Camera can output composite video and audio directly to a video monitor or TV, or digital data can be output on a special serial cable. This can be connected to a PC using a plug-in board that interfaces directly with the computer's system bus. In addition, data can be transferred by removing the hard disk card from the camera and plugging it into a computer that has a PCMCIA Type III port.
- It has an LCD viewfinder that also is used to view playback of stills or motion video from the hard disk. A directory of the hard disk can also be viewed from the LCD. Any file on the disk can be selected for playback or deletion.
- There is a 3× optical zoom and an additiona 2× digital zoom. Since the CCD has twice the resolution of the compressed motion video, the digital zoom should not cause any loss of resolution in video. For stills, however, the digital zoom does cause loss of resolution.
- The camera operates from rechargeable batteries. An ac adaptor/charger is provided for powering directly from ac or charging the batteries.
- Bundled software is provided with the product for interfacing to a computer for storage and editing of audio, video, and still images.
- A remote control unit allows operation of most camera functions via infrared communication.
- Automatic exposure and white balancing are provided.

The Hitachi MPEG Camera is the first of a new breed of products that are blurring the line between digital still cameras and traditional video camcorders.

8.4 CAMCORDER DESIGN

There are many mechanical engineering considerations involved in designing a lightweight camcorder that will be inexpensive, while high-performing, rugged, reliable, and easy to use. These are discussed in this section.

8.4.1 Ergonomics

The considerations of how to make a product physically suitable for its application and easy to use are often grouped under the heading of *ergonomics*.

8.4.1.1 Configuration

There are two basic camcorder configurations, with many variations—shoulder-held and hand-held. Early camcorders were so heavy that the only way one could hold them for any length of time was by sitting them on the shoulder. As technology advances made smaller camcorders possible, hand-held models were introduced. However, the hand-helds did not displace shoulder-held models for several reasons.

It should be pointed out that all camcorders have provision for tripod mounting and this is the preferred mode in all situations to get the most stable pictures. However, the fact that a camcorder can be taken off the tripod and carried around to get shots that would be impossible with a tripod is a camcorder's greatest advantage over fixed cameras.

Shoulder-mounting is the most stable and safest way to carry a camcorder. The camera is supported and protected by the upper body, which also provides the smoothest motion for the camera. There is a minimum size for a shoulder-held camera regardless of technology; that is caused by the distance between a person's shoulder and his or her eye, which must be able to see the viewfinder and, by the need to make the camera balance on the shoulder. Thus, there is plenty of room in a shoulder-held camcorder for large tape formats, elaborate control panels, and so on. Shoulder-held is the preferred configuration for semi-professional and professional uses.

A hand-held camcorder is held in front of the operator's face, in a position for convenient use of the viewfinder, which is on the rear of the unit. The camera weight needs to be low because the operator's hands are supporting the entire camera weight. This is a less stable mode than shoulder-mounting and most professional camera operators prefer the latter. Also, light weight inherently means small size, so there is not as much room on a hand-held camera for convenient controls and displays. Hand-helds are preferred in the home market probably because they are small, convenient, and inexpensive.

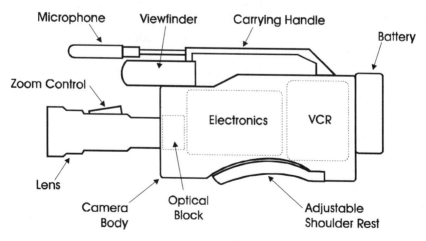

Figure 8.4 *Typical configuration of a shoulder-held camcorder.*

8.4.1.1 Shoulder-held Camcorder Layout

Shoulder-held camcorders are designed to balance on the operator's shoulder. This, in combination with the eye-to-shoulder distance constraint, leads to a more or less standard arrangement of components, as shown in Figure 8.4. The heaviest elements of a camcorder are the battery, VCR, and lens. These balance each other over the shoulder as shown in the figure. The back-and-forth position of the shoulder rest generally has to be adjustable to maintain balance when different types of lenses or batteries are used. The viewfinder usually extends beyond the front of the camera and, a microphone, if used, extends even further out to get away from the noises of the camera mechanisms.

In operation, the camera operator carries the camera on his or her right shoulder with the right eye at the viewfinder eyepiece. The camera body may rest against the head for stability; some cameras have a soft pad on their left side to make this more comfortable. The right hand holds the lens and operates the lens controls for zoom and focus (if not using automatic). A strap is often provided on the lens control unit to hold the operator's hand; that keeps the camera from getting away from the operator in a tight situation. The camera is moved about by rotating the operator's upper body, keeping the camera steady on the shoulder. The operator's left hand is free for manipulating controls, pushing through crowds, providing additional support to the camera, and so forth. Some camcorders are adaptable for operation on either the right or left shoulder, although this is unusual.

8.4.1.2 Hand-Held Camcorders

Hand-held camcorders have few physical constraints on their design other than weight and, therefore, there are many more variations of configuration. Figure 8.5 shows some of these. Two categories of use are generated by the viewfinder type: an eyepiece viewfinder requires the camcorder to be held against the operator's eye, whereas a direct-view viewfinder must be held away from the head far enough for the eye to focus. Of course, the latter mode gives much more flexibility in holding the camera; it can be held up or down and the viewfinder swiveled so it can be seen. The disadvantage of this is that the camera cannot be steadied by holding it against the operator's head. Many hand-held camcorders have addressed this problem by including image stabilization.

The ¼-in tape DVD digital format is a boon to hand-helds because it has the smallest tape transport of all. The Sony DCR-VX1000, the JVC GR-DVM1, and the Panasonic AG-EZ1U all enjoy this advantage.

8.4.1.3 Operating Controls

There are three categories of controls needed on a camcorder:

- Operating controls—these are controls used continuously during shooting, which include the zoom, focus, and iris controls for the lens (unless some are automated,) and the VCR Record control. These controls are generally located so they are easily operated while the camera is being held in the normal shooting position. On

Sony CCD-TR96

Sony DCR-VX1000

Sharp VL-E650U/VL-E620U

JVC GR-DVM1

Panasonic AG-EZ1U

Figure 8.5 *Typical hand-held camcorders.*

shoulder-held camcorders, the normal shooting position is with the camera sitting on the shoulder and steadied by one hand around the lens. The operating controls are on the lens close to the holding hand. Similarly, hand-held cameras have their operating controls (usually only the zoom and record controls) positioned where the user's hand can operate them while holding the camera.

- Mode-change controls—these include on-off switches for automatic functions, video gain switching, white balance switches, and switches that change the viewfinder presentation. These are used occasionally during operation and should be located so they can be accessed without interrupting shooting. On a shoulder-held camcorder, the lower left side and lower front of the camera below the lens are the preferred location. These can be reached by the operator's left hand without moving the camera. The switches should be designed so they can be operated by feel because this location is too close to the operator's head for he or she to be able to see them without moving the camera. The best location on hand-held cameras for mode switches varies with the configuration, but they should be in easy reach of the hands while holding the camera and should be recognizable by feel.

- Setup controls—these are other controls that would not normally be used during shooting. The camera is generally taken out of shooting position (off the shoulder or down from the eye) to operate these controls. They include such things as tape speed, tape loading, video and audio setup, tape preview and editing, and so forth.

Another type of "control" is a status display. Most camcorders have LCD panels for this purpose and for use during setup procedures. These displays are generally not practical to use during shooting and any status information needed while shooting should be presented in the viewfinder.

8.4.1.4 Audio Considerations

Although an on-camera microphone will never match the performance of an external mike placed in the scene or held by a performer, there are many situations where the on-camera mike is the only practical approach. Thus, the camera designer should do as much as possible to get the best results from the on-camera mike. That means mounting it as far forward as possible, using a directional pattern to reduce pickup from the camera and operator, and reducing the noise generation within the camera.

Acoustic noise sources in a camcorder include the lens, control switches, and the VCR. Sounds from these sources picked up by the mike become a part of the program audio just like sounds from the scene. Good camera design will take steps to reduce the noises coming from the camera.

8.4.1.5 Battery Mounting

Batteries, of course, must be removeable for recharging. Most camcorder batteries are mounted at the rear of the unit, away from the lens, where the battery weight helps balance the lens weight at the other end. Some form of snap mounting is used to simultaneously hold the battery in place and make the electrical connections. Often, there are several sizes of battery that can be used; these are all accomodated with the same mounting. If the camcorder provides a battery life indicator, the right place for that is in the viewfinder.

The general subject of batteries is covered in Section 10.3.1. In addition to batteries supplied by camera manufacturers, there is an active after-market of specialty battery companies that often offer improved performance or convenience in battery applications.

8.4.1.6 Loading of Tape

The tape cassette for the VCR is generally placed into a some kind of slot and a door is closed manually or electrically to activate the transport threading process. To remove the tape, a button is pressed, activating the unthread process, following which the door opens and the cassette is brought to a position where the user can easily retrieve it. It is desirable for tape loading and unloading to be done without undue care being necessary. On a shoulder-held camcorder, it is ideal if tapes can be changed without lifting the camera off the shoulder.

Of course, it is obvious that the design should fully protect the mechanisms from incorrect tape loading actions. Some camcorders have not followed this rule and a tape loading error could send the camcorder to the shop.

8.4.1.7 Cable Connections

Cables running to any portable unit are a liability. They can become tangled, catch on things, break at stress points, and so on. Cable connections to a camcorder should be positioned so they do not interfere with operation and so proper stress relief can be given to the cable. On shoulder-held camcorders, the lower rear seems to be the best position and most units have some type of recessed panel there for external connections. Cables connecting at this position can naturally drape over the operator's back. It is usually a good idea to provide means for anchoring such cables to the operator's belt so a snag on the cable will not stress the camcorder connections or trip the camera operator.

8.4.1.8 Environmental Considerations

Field equipment faces a wide range of environmental conditions: temperature, humidity, atmospheric pressure, vibration, weather extremes, and so forth. Camcorders are inherently delicate instruments and there are limits in their capability to withstand these stresses, even with the best possible design. Camcorder specifications generally say what the acceptable extremes are for temperature and humidity, but limits for other stresses are often not specified. A camcorder should have a degree of weather-tightness so it can stand being caught out in a light rain, but more extreme conditions require external protection for the camcorder.

The ultimate environmental catastrophe for a camcorder is for it to be dropped. Almost any drop of a distance more than a few inches will cause some damage, especially to the lens. The camera body strength is limited by weight requirements, but most camcorders do offer some protection to the electronics and the VCR. However, the VCR has many points of extreme mechanical precision and it is easy to see how even a moderate drop will cause something to go out of alignment. Other vulnerable points are viewfinder and microphone mounts. The best protection is for camera operators to be careful.

8.4.2 Dockable Camcorders

One variation on the shoulder-held camcorder is the *dockable* configuration. This simply means that the camera and the recorder can be separated. By replacing the recorder of the camcorder by an interface unit, the camera can be used by itself in a studio setup. Similarly, it is conceivable that the same camera could be used with different recorders for different tape formats. These features are available is some product lines, but they seem to be becoming less popular as the basic cost of camcorders goes down. It is generally more convenient to have complete camcorders for each required configuration. The small cost saving of dockability is offset by inconvenience in operation.

8.4.3 Camcorder Viewfinders

Chapter 9 gives a complete story on viewfinders; this section covers only the issues specific to camcorders. Two categories of camcorder viewfinders are eyepiece and natural-vision viewfinders.

8.4.3.1 Eyepiece Viewfinders

The traditional camcorder viewfinder uses an eyepiece that the user puts against his or her eye to view a video display, usually a small CRT or color LCD, within the viewfinder. Since the video screen is very close to the eye, a lens is required so the eye can focus so closely. The advantages of this type of viewfinder are that it can be quite small and a flexible hood on the eyepiece can shield out natural light so the display does not have to be too bright. The disadvantage is that the user's eye is trapped by the viewfinder. It is difficult to look around the scene with the viewfinder eye; the camera operator has to learn how to do that with his or her other eye.

On shoulder-held camcorders, the viewfinder is usually the most forward unit on the camcorder except the lens (see Figure 8.4). Because a CRT display has significant length behind the screen, placing the CRT directly in front of the eyepiece would make the forward extension even greater. Most camcorders using CRT displays solve this problem by placing the CRT crosswise directly in front of the eyepiece and using a mirror or prism to make a right-angle turn in the optical path to the eyepiece. That brings the viewfinder housing in closer to the camera housing where it is less vulnerable. Of course, a color LCD, being smaller, simplifies some of the design considerations.

On hand-held camcorders, the viewfinder is usually located at the top of the camera body and there is enough length of the body for the CRT to be in the direct-viewing position. Some camcorders provide for the viewfinder to rotate upward from the camera body, which is convenient in many shooting situations.

8.4.3.2 Natural-Vision Viewfinders

A larger-sized LCD panel of 2 to 4 in diagonal can make a viewfinder that does not require the eye to be trapped in an eyepiece. The LCD panel is swivel-mounted so that it can be placed at any angle to the camera, resulting in a viewfinder that the operator can see regardless of the position in which he or she is holding the camera. LCDs are generally in color.

Because of these advantages, LCD-panel viewfinders are offered on many hand-held cameras, either as an option or, sometimes, as the only choice. They are convenient and, for their image size, smaller than CRT viewfinders. However, there are some disadvantages: LCD panels may be difficult to see in bright ambient light and, of course, they require holding the camera far enough away from the eye for focusing. This usually leads to an "arms-extended" posture for the operator, which is fatiguing and vulnerable.

Direct-view LCD viewfinders are not too applicable to shoulder-held camcorders because of the eye-to-screen distance required for focusing. Of course, a lens could be used to help the focusing, but that is undoing the advantage of direct viewing. Direct-view LCDs will probably continue to be an option that attracts some users but they are not for everyone.

8.5 CAMCORDER FEATURES

Camcorders have evolved quite a large list of feature possibilities. Some are features seen in all types of cameras but others are specific to camcorders. Only the latter features are discussed here.

8.5.1 Automatic Operation

A camcorder operator is responsible alone for what goes onto his or her tape while shooting. No video operator back in a control room somewhere is tweaking signals as he or she is shooting. Because there are many other concerns for a camcorder operator while shooting, there is little time for technical considerations and the best choice is that all technical parameters should be fixed or automatic. The camcorder operator has a full job just framing the shots and maintaining his or her position at the scene.

Automatic features for white and black balancing, exposure, and focusing were discussed in detail in Chapter 5. These features are most important to camcorders. However, there are situations where even the best automatic system does not do what is required with the picture. There may be a flickering highlight that upsets the automatic exposure system, or a severe focus compromise between different parts of the scene may be impossible for the automatic focus to handle. In these cases, a skilled operator wants to shut off the automatic feature and take over manual control. Camcorders should have such switches and also provide easy-to-operate manual controls for those cases where they are needed.

8.5.2 In-Camera Editing

In professional production, editing is done in a postproduction studio with elaborate facilities and with all the people responsible for the program in attendance. This cannot be done with a camcorder. However, a home user does not have such facilities but he or she would still like to accomplish some of the things they do in professional production (see Section 6.4.3). Thus, home camcorders have extensive editing features but professional ones do not. Professional camcorders usually do have basic add-on editing simply because that facilitates building a multishot reel without the interruptions and loss of synchronization that would otherwise occur on every stop and restart of the tape.

Another editing-related feature in some professional digital camcorders is the ability to add additional data to the video tape that will assist in later editing of the tape. The most important of these is time code, which all professional camcorders provide and it is even beginning to appear in home camcorders. Time code allows specific points in the video to be marked, and shots identified. Some camcorders also allow comment text to be added to the video. When the tape is played back for entry into a nonlinear editing system, this information is displayed and can even be used to control what shots are entered into the system and what shots may be skipped. This speeds up the editing process.

8.5.2.1 Editing with a Home Camcorder

High-end home camcorders have the capability of limited editing in the field, using the tape video as the first input and camera video as the second input. More elaborate editing can be done using the camcorder as one of the VCRs in an edit controller setup. This calls for remote control capability on the camcorder.

Editing requires replaying recordings to choose the in-edit point. On a home camcorder without time code, the in-edit point is determined by playing back and stopping the tape at the desired frame. Depending on the type of VCR controls on the camcorder, this may be easy or difficult. Upon starting an edit, the camcorder will automatically rewind the tape an appropriate amount for run-up and then begin playing. When the desired edit in-point is reached, recording is automatically switched on (for a cut transition). Some camcorders can do certain other transitions, such as fades or wipes. These are generally done by saving a still frame from the outgoing video to mix with the live incoming video during the transition. The operator must manually tell the camcorder when to stop recording. In-camera editing requires a lot of skill to get the timing right and to coordinate the live scene with the operation of the VCR.

Besides the timing problems mentioned above, a home camcorder has the problem that there is usually not room for a convenient control panel on the small camcorder package; editing controls are likely to be located all over the camera, which also makes the task more difficult. It helps to use remote control, and to add an external video monitor that can be viewed regardless of the position of the camcorder. But for serious editing with home equipment, the user should consider real postproduction using a second VCR and an edit controller.

8.5.3 Video Output

The usual output from a camcorder is the tape. Video is viewable at the camcorder only in the viewfinder. However, most home camcorders provide video and audio outputs so the camcorder's video and audio can also be viewed on a TV receiver. This is especially necessary if the camcorder is the only tape playback device available.

The other use for camcorder video outputs is to connect the camcorder into a multicamera system. Professional camcorders are generally adaptable to simple systems, where a degree of central artistic control is desired. Video and audio signals go to a central location where they are viewed and the camera operators can be directed. However, most camcorders do not have intercom capability, so this has to be set up separately for such systems.

High-end professional camcorders may have the option of a triax interface to a large system, giving them access to nearly all the features of high-end studio cameras with the plus of having a local VCR.

Another feature of some digital camcorders is an option to play back the tape at double speed in the camcorder and deliver a digital output that can be transmitted to a server for rapid loading of a nonlinear editing system. This feature in in Sony's Betacam SX camcorders.

8.5.4 Remote Control

Remote control of camcorder operating functions is useful in many situations. If the person operating the camera wants to become part of the scene, it is best done by having remote control of the camcorder. That occurs primarily in home shooting because, in a professional environment, the people operating the equipment are usually different from those who perform in front of the lens.

Home remote control units generally use infrared communication with the camera. A typical remote controller will have a zoom control and VCR record, stop, rewind, and play controls.

Professional remote controllers are generally wire-connected. That provides greater reliability and distance capability. Many other control functions may be included, such as video levels, manual iris, gain controls, mode switches, and so forth.

8.5.5 Memory Card Setup

Some professional camcorders have a memory card feature similar to that found in professional studio cameras. Setup parameters are stored on a removable memory card at the camcorder. Just as in studio cameras, this feature allows multiple cameras to be given the same setup or for a single camera to be quickly changed from one setup to another.

8.5.6 Time Code

Professional operations require that all tapes be recorded with time code. Thus, professional camcorders generally have time code generators built-in. Since it is useful to log time codes of scene starts and endings during shooting, the time code generator should have long-term battery backup so it can be set precisely to clock time. That way, anyone on the scene can log times from his or her own watch.

Time code is seldom used in home camcorders. However, as the market for home editing grows and users begin to build multi-VCR editing suites, there will be more demand for time code. It is beginning to be seen in high-end home camcorders.

8.5.7 Titling

In professional production, titles are generated and inserted in postproduction and there is no use for them in a camcorder. However, in a home camcorder, a titler is a nice plus feature. With digital ICs, almost any degree of titling sophistication could be built in a small, low-power package. That is not the problem—the problem is *controlling* a fancy titler. An alphanumeric keyboard, many selectors, and a display is generally required on a free-standing titling unit; there's no room for all that on a small home camcorder.

Home camcorder titling usually has some form of letter-at-a-time input device, where buttons select each letter from a loop of all possible letters. This is an awkward way to input text and limits the systems to a small number of total characters. Similarly, the choices for text size, styles, and colors is limited by the controls available. Some camcorders have a memory that can store multiple title screens; when a title is inserted, the desired one is selected from the memory. Again, that takes controls, so the flexibility is limited.

8.5.8 Genlock Input

When a camcorder operates by itself in the field, there is no need for synchronizing anything. However, when the same camera units are used in a multicamera setup, as described above, the cameras should be synchronized. That is accomplished by sending a video signal out to each camera for it to synchronize with. This is called *genlocking*. Most professional camcorders have this feature.

REFERENCES

1. Inglis, A.F., and A.C. Luther, *Video Engineering, Second Edition*, McGraw-Hill, New York, 1996, Chapters 9 and 10.

2. Luther, A.C., *Principles of Digital Audio and Video*, Artech House, Norwood, MA, 1997, Chapter 11.

3. Ibid.

4. Ibid., Chapter 7.

9

Viewfinders

The electronic viewfinders used on virtually all video cameras offer many features to make camera operation easier. This chapter discusses viewfinder design and use.

9.1 VIEWFINDER REQUIREMENTS

A camera viewfinder does a lot more than just show the operator the picture from the camera. This section lists the characteristics of electronic viewfinders.

9.1.1 Viewfinder Features

Electronic viewfinders are very different from the optical viewfinders used in photographic cameras. Some of their different features are listed here:

- The viewfinder is not tied to the camera optics and therefore can be located anywhere on the camera or even away from the camera.
- Like the optical finder on a *single-lens reflex* (SLR) photographic camera, the electronic viewfinder shows exactly what the output picture will be. In computer terms, this is *WYSIWYG* (meaning "what you see is what you get," pronounced "wissy-wig").
- There can be more than one viewfinder, even of different types.
- The viewfinder can be any size, from so small that it must be viewed with a magnifier up to TV-screen size or even projection.
- The viewfinder image can be enhanced to highlight image properties that are important to camera operation; for example, image sharpness can be enhanced to assist focusing.
- The viewfinder can show things other than or in addition to the camera's own picture. Examples are data displays, video from another source, or cursors marking specific points or areas of the screen.

Table 9.1

Advantages and Disadvantages of Viewfinder Configurations

Type	Advantages	Disadvantages
Eyepiece	User can see a relatively dim viewfinder in bright ambient illumination. Viewfinder can be small.	User's eye is trapped in the viewfinder, making it difficult to view the scene directly.
Direct view	User can easily view the picture and the scene directly. User can move about independently from the camera.	The user must be far enough from the viewfinder for his or her eyes to focus on the screen. Viewfinder must have similar brightness as the scene.

These features are further discussed in the rest of this chapter.

9.1.2 Viewfinder Configurations

There are two types of viewfinders—eyepiece viewfinders where the user looks into an optical system that allows close viewing of a small display device, and direct-view viewfinders where the user looks normally at a screen without any intervening optics. These have advantages and disadvantages as listed in Table 9.1.

9.1.3 Viewfinder Performance

The key performance specifications for camera viewfinders are listed here:

- Image size—This ranges from the small sizes used for eyepiece viewfinders, with diagonals from about 0.5 to 1.5 in, to larger sizes for direct viewing, whose diagonals are in the range of 4 to 7 in.
- Resolution—Ideally, a viewfinder should have as much or more resolution capability as the camera it accompanies. However, this is often impractical and viewfinders may have less resolution than their cameras and still be successful. The reason for high resolution in a viewfinder is to be able to see when the camera is precisely in focus. Obviously, this may be unnecessary in cameras that have automatic focusing and, for those cameras, only enough resolution is required to display the picture for framing the image. A resolution of 200 to 300 TVL is adequate for that purpose. However, for manual focusing, viewfinder resolutions up to 600 TVL may be necessary. To some extent, a lack of resolution can be compensated by enhancing the image to exaggerate focusing effects.
- Monochrome or color—Viewfinders have traditionally been monochrome but, with the advent of color LCDs, color viewfinders are appearing on home cameras.

However, these have limited resolution and are suitable only for automatic focusing. Professional cameras still mostly have monochrome viewfinders because they can deliver the higher resolutions that are required for manual focusing. Professional camera manufacturers offering optional color viewfinders have not sold very many.

- Brightness—A camera can operate in a wide range of illuminations, ranging from full sunlight to near darkness. The viewfinder must be visible under all these conditions. Eyepiece viewfinders inherently are shielded from the ambient illumination by the eyepiece cup; however, the viewfinder brightness cannot be too dim because the operator is looking at it with only one eye and he or she must use the other eye to watch the scene directly to be aware of things happening outside the view of the camera. Too much difference in brightness between the scene and viewfinder means that the operator's eye must adapt to brightness every time he or she looks directly at the scene.

However, the brightness situation is more difficult for direct-view viewfinders because there, both of the operator's eyes are adapted to the scene brightness, and the viewfinder brightness must be in the same range.

- Mounting—the viewfinder mounting must be highly adjustable to accommodate all the positions that may be taken by camera and operator. For example, a studio camera viewfinder should be conveniently viewable by the operator as the camera is tilted up or down or raised or lowered. An example of this is shown in Figure 9.1, where the viewfinder can be raised up and rotated about a horizontal axis for viewing when the camera height is different from the operator's eye level.

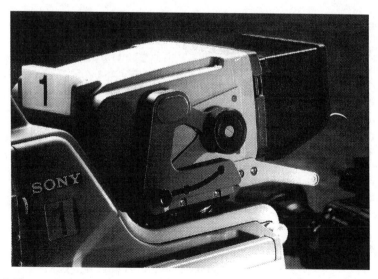

Figure 9.1 The viewfinder mount of the Sony BVP-700 studio camera. (Courtesy of Sony.)

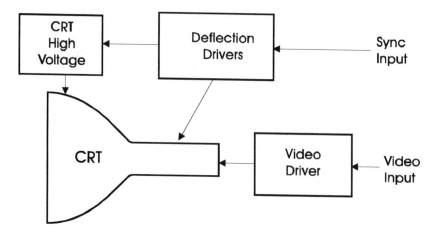

Figure 9.2 *Block diagram of a typical CRT viewfinder.*

9.2 CRT VIEWFINDERS

For many years the CRT was the video display device of choice simply because there was no other choice. However, in the last ten years, other devices, such as liquid-crystal displays (LCDs), plasma display panels (PDPs), digital micromirror devices (DMD), and other technologies have established positions in the display market along with the CRT. Only two of these, the CRT and the LCD have seen much use as viewfinder devices. This section discusses the application of CRTs to viewfinders.

9.2.1 The Cathode-Ray Tube

For viewfinder applications, CRTs provide high brightness and high resolution. However, being vacuum tubes, they suffer from the inherent limitations of such devices: limited life, slow warmup time, requirement for high voltages, sensitivity to magnetic fields, and performance drift with time, temperature, and other factors. Nontheless, CRT viewfinders are widely used, especially in professional cameras.

Figure 9.2 shows a block diagram for a typical CRT viewfinder. Magnetic-deflection CRTs are used and the circuits are similar to those in TV receivers except that the power level is less because of the smaller tubes. However, the need for high brightness and high resolution in direct-view viewfinders calls for relatively high anode voltages on the CRT.

9.3 LCD VIEWFINDERS

In many respects, a *liquid-crystal display* (LCD) is an ideal viewfinder device. It is small, low in power, does not require high voltages, and has good reliability with long life. LCDs are available in both color and monochrome versions, in sizes from 0.5 in up, and with resolutions up to 600 TVL.

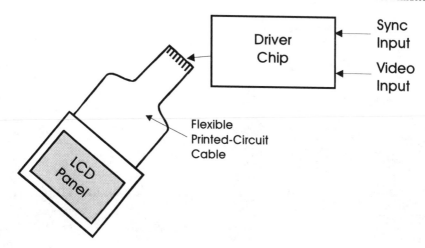

Figure 9.3 *Block diagram of a typical LCD viewfinder.*

9.3.1 LCDs

The LCD is based on the electronic properties of a class of materials known as liquid crystals. These materials are called *nematic*, because they have elongated rod-shaped molecules. When placed in a container whose interior surfaces are slightly grooved, the LC molecules will naturally align themselves with the pattern of the grooves. When an electric field is applied to this structure, the molecules will shift themselves to more or less align with the electric field. By passing polarized light through the LC container, the plane of polarization is shifted by the alignment of the molecules. When viewing the light from the LC through another polarizer of fixed angle, the light transmission is modulated by the amount of rotation of polarization caused by the magnitude of the electric field placed on the crystal. By building the crystal with an orthogonal structure of electrodes on each surface, light transmission is controlled on a pixel basis according to the voltages on the electrodes [1].

9.3.2 LCD Applications

Figure 9.3 shows a block diagram of an LCD viewfinder. Most LCDs are available with a companion driver chip that handles the scanning and video driving operations, so their application is very simple. The LCD itself is connected by a flexible printed circuit cable, which might be wrapped around behind the LCD to mount the device in a narrow housing in the case of an eyepiece viewfinder. The device is mounted in the viewfinder housing and a suitable source of rear illumination provided. In some viewfinders, the illumination is provided by a window that lets in ambient light behind the LCD. Of course, an electrical illumination source is still needed for use when the ambient light level is too low.

(a) Out of Focus   (b) Focused

Figure 9.4 *Focusing by using excess image edge enhancement. (Courtesy of Sony.)*

9.4 VIEWFINDER SIGNAL PROCESSING

The effectiveness of a viewfinder can be improved by special processing of the video signal.

9.4.1 Image Enhancement

Focusing to a very high precision can be done using a viewfinder that has much less resolution than the camera itself if the edge transitions in the viewfinder video are artificially enhanced. That is done by the same process as aperture correction (see Section 4.5), using digital high-pass filters to generate horizontal and vertical enhancement signals. By applying an excess amount of these signals, all edges in a properly focused image will show up as severely overcorrected, which produces a very visible outline effect. In viewfinders, this type of enhancement is called *peaking*.

For peaking to work even on images that do not have a lot of sharp edges, the frequency range of peaking is set lower than that used for aperture correction. Since the visibility of peaking reduces as the image goes out of focus, one simply focuses for the maximum amount of overcorrection. This is an operator-preference item and most viewfinders provide an adjustment so the operator can set the amount of peaking that satisfies him or her. Peaking is very effective. Figure 9.4 shows the effect on a typical image. The white outlines on the pencils occur only when they are in sharp focus.

| Safe Area Cursor | Center Marker | Cross-Hair Cursor |

Figure 9.5 *Viewfinder cursors.*

9.4.2 Cursors

Viewfinders are usually scanned so that the entire picture is visible. However, it is often desirable to show the camera operator a specific area of the picture within which he or she should keep the principal action of the scene. This is often referred to as the *safe action area* and, in television, it represents the area of the picture that will be visible on a typical overscanned TV receiver.

Viewfinders can display a line rectangle (*cursor*) showing the safe action area. This is usually provided as a switchable option in the viewfinder and, in some cameras, the dimensions of the rectangle are programmable to any values. That allows the cursor to serve other purposes, such as showing an area of the picture that will be replaced by another signal in postproduction.

Another example is a cursor that shows the center of the screen. That is usually in the form of a cross marker and it is used by the operator for precise centering of objects in the picture when that is called for. The existence of and sometimes the position of the cross cursor is programmable in many cameras. Sometimes this is in the form of a cross-hair that can be adjusted either locally or remotely to identify a specific screen location. Figure 9.5 shows some examples of viewfinder cursors.

9.4.3 Return Video Display

When a camera is used in a system that is combining multiple camera signals in real time, it is often useful for the camera operator to see how his or her signal is being mixed with other signals. That way, the operator can make adjustments in camera positioning or framing to achieve the desired effect when the signals are combined. Most system cameras have a *return video* feature for this purpose.

Ordinarily, it is not satisfactory to simply display the return video signal full-screen on the viewfinder because this limits how much of the camera's own signal can be seen by the operator. Cameras provide various features for displaying the return video, including splitting the screen, displaying it (or the camera video) in a window (*picture-in-picture*), or simply switching between the camera signal and the return video. Figure 9.6 shows some examples of return video displays for an example where a commentator in front of the camera has another picture (the biplane) inserted in the background over his shoulder.

Camera Signal   Return Video

Picture-in-Picture   Picture-in-Picture
                             (Signals Exchanged)

Figure 9.6 *Viewfinder return-video displays.*

9.4.4 Highlight Processing

When a video camera operates with automatic exposure control, the operator does not need to worry about video level but, when the camera is in manual exposure mode, there needs to be a way for the operator to know when the video level is correct. This is done in the viewfinder by modifying the display as a function of level so the operator can see where the highest levels are in the picture and how they are set with respect to reference white level. The usual approach for this is to superimpose a pattern of stripes over the highest-level areas of the viewfinder picture. This is called *zebra* processing.

It is most effective to use two levels of zebra processing, as shown with a gray scale pattern (see Section 12.2.2) in Figure 9.7. The first level of zebra appears when the video level is close to 70% of reference white. This is used to set exposure of skin tones in the scene for a video level of 70%, which usually gives the most pleasing skin reproduction. The second-level (coarser) zebra pattern appears for scene areas that will go above reference white level and may be clipped later in the system. Generally both zebra settings are adjustable by the user to achieve the desired result. Because the zebra pattern interferes with the normal use of the viewfinder, it is usually controlled by a momentary switch that is turned on only when setting video level.

9.4.5 Data Displays

There is a lot of information that the camera operator may need to know about what is going on inside his or her camera. Since the operator looks at the viewfinder all the time while shooting, it is the natural place for displaying such information.

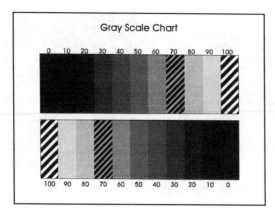

Figure 9.7 *Viewfinder zebra display to show video level.*

Examples of data display parameters are:
- Battery condition (camcorders);
- Automatic/manual control choices;
- Iris setting;
- Gain setting;
- VCR mode (camcorders);
- Tape remaining (camcorders);
- Zoom setting;
- Camera ID;
- On-air tally;
- Filter wheel position(s);
- Electronic shutter mode;
- Fault warnings;
- Text messages sent from the camera control position;
- Setup mode menus.

Obviously, not every camera has all these features in its viewfinder; there has to be room left to display the picture along with the data. Some examples of viewfinder data display are shown in Figure 9.8.

9.5 CONCLUSION

The viewfinder is the camera operator's "eye" into the picture produced by his or her camera. As such, it it is important in assisting the operator to achieve the best possible artistic and technical quality in the picture.

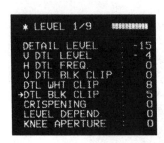 

Figure 9.8 Examples of viewfinder data displays. (Courtesy of Sony.)

REFERENCE

1. Luther, A. C., *Principles of Digital Audio and Video*, Artech House, Norwood, MA, 1997, pp. 257–259.

10

Camera Packaging

The aspect of design where all the components and features of a camera are assembled into a unit that fulfils all the objectives for the system is its *packaging*. This embodies a very large measure of mechanical engineering although that is definitely not the whole story. Factors of human engineering, thermal design, manufacturability, and other disciplines also are involved. Many packaging considerations have already been mentioned in the previous chapters, but this chapter covers packaging as a whole.

10.1 PACKAGING REQUIREMENTS

Packaging embraces just about everything about a camera except the optical design and the electronic circuit design and, even there, many of the considerations in the following list apply:

- Ergonomics;
- Size and weight;
- Physical stability;
- Ruggedness;
- Flexibility of mounting and use;
- Environmental protection;
- Thermal considerations;
- Interfaces;
- Manufacturability.

These factors are discussed further in the following sections.

10.1.1 Ergonomics

Probably the one packaging factor that contributes most to the success of a camera design is *ergonomics*, which is the design for usability and ease of use. Whole books have been written about this subject, which is sometimes called *human engineering*. This section discusses the most important ergonomic factors in cameras.

10.1.1.1 Camera Handling

Whether a hand-held camcorder or a studio camera mounted on a pedestal, crane, or dolly, the first consideration about handling is that the operator must easily be able to produce smooth camera movement and that it be convenient and comfortable to do over long periods of operation.

For hand-held cameras this requirement calls for a hand-holding strategy that provides a balanced and stable configuration that also gives convenient access to the operating controls while the camera is in use. Because of the latter, hand-held cameras are usually designed for one of the operator's hands to hold the camera body so that the fingers tips can reach an operating controls panel located convenient to the hand-hold. In many cameras, the use of the other hand is optional; if used, it holds the camera at the rear or bottom to provide more stability. The hand-hold often includes a strap around the hand so that the camera can be better held in one hand without risk of dropping it.

With shoulder-held cameras the camera sits on the operator's shoulder and the right hand holds the lens; in this case, the shoulder carries the camera weight and the hand on the lens controls stability. Operating controls are positioned on the lens where they can be reached by the hand holding the lens. The operator's other hand is free to operate controls, push through crowds, and so on.

A tripod- or pedestal-mounted studio camera does not need to be held by the camera operator.It should be easily moved by the operating handles but it should remain in position when the operator releases the handles. This capability is an objective of the tripod head (see Section 10.3.1.1).

10.1.1.2 Visibility Around the Camera

A camera operator spends most of his or her time looking into or at the viewfinder. However, at all times it should be easy for the operator to glance beyond the camera to the scene. That is necessary so the operator can stay aware of what may be happening outside the scene area currently being captured by the camera or to look for potential hazards that may limit movement of the camera or the operator's position. This is difficult with an eyepiece-viewfinder camera; the operator's eye in the viewfinder is not easily moved away and back again to see the scene. Most operators learn to use their other eye for looking around. This is still a consideration of camera design because the camera should block the vision of the other eye as little as possible. The most important requirement here is that the camera is not too high, which would make it difficult to see over when the operator needs to view the scene on the other side of the camera.

Figure 10.1 *The Sony BVP-700 studio camera. (Courtesy of Sony.)*

With a tripod-mounted camera having a direct-view viewfinder, the operator stands back from the camera far enough for him or her to comfortably focus on the viewfinder. In this case, the camera and viewfinder should not cause a large block in the operator's straight-ahead vision.

A second consideration in studio cameras is that the camera movements are most intuitive for the operator if the viewfinder is at the same level as the lens. That way, the lens sees the scene the same way the operator would see it if the camera were not there. This has led to the popular studio camera design where the rear of the camera body is lowered so the viewfinder can sit almost at the same level as the lens. The Sony BVP-700 camera shown in Figure 10.1 is an example of such a design.

10.1.1.3 Operating Controls

The lens zoom control is the one control that the operator uses continuously on all cameras. When automatic focusing or automatic iris is not being used, the operator must also continually have access to these controls. In a camcorder, the other "operating" control is the VCR record switch. In all cameras, these controls must be located so the operator can use them without moving his camera-holding hand, which might interfere with camera motion or stability.

With hand-held cameras, the operating controls are placed convenient to the hand that is holding. In tripod- or pedestal-mounted cameras, operating controls are usually

Figure 10.2 *Lens-mounted operating controls on the Sony Digital Betacam.™ (Courtesy of Sony.)*

placed on the camera-holding handles. This configuration can be seen in Figure 10.1. Some cameras provide rotation of parts of the handle for control; others mount various types of actuators or cranks to the operating handle. This is very much a matter of operator preference and many studio camera product lines provide a choice of the control methods that can be selected.

Since most of the controls interface with the lens, lens manufacturers provide operating controls that can be adapted to any camera using their lenses. The method of extending control from the lens to the handle area may be either mechanical or electrical. Mechanical systems usually use flexible-shaft transmission; electrical methods include motor speed controls, position servos, or even computer control.

10.1.1.4 Secondary Controls

Controls that are used occasionally during shooting are secondary controls. These include such things as white balancing, mode changing (e.g., auto on-off functions or gain switches), viewfinder setup and mode controls and, on camcorders, VCR setup and mode controls. Generally, the use of secondary controls occurs during breaks in the shooting and the operator can shift his or her attention from the shot to operate them. These controls should be located within the operator's reach without excessive movement and without opening doors. It is desirable that they be locatable by feel so they can be operated while the operator still watches the viewfinder.

10.1.1.5 Setup Controls

The remaining controls on cameras are for setup or diagnostic procedures, or for special processes such as editing on a camcorder. Setup generally is done away from the shooting

Table 10.1
Camera Size and Weight

| Type of Use | Objectives | |
	Size	Weight (lb)
Hand-held	As small as possible	1 to 5
Shoulder-held	At least large enough for shoulder-to-eye distance	5 to 15
Studio (tripod or pedestal)	Large enough to support lens and viewfinder of intended sizes	20 to 60

situation. Such controls can be behind doors and may require the operator to move away from the usual operating position of the camera.

10.1.2 Size and Weight

The packaging design of a camera is obviously substantially affected by the objectives for size and weight of the camera. In turn, these objectives depend on the intended use for the camera. A hand-held camera should not be too heavy or too large; a shoulder-held camera must be large enough to sit comfortably on the shoulder while the operator watches the viewfinder, a studio camera must be large enough to support the intended lenses, and so forth. These factors are summarized in Table 10.1.

Whether hand- or shoulder-held, weight does add stability to the camera. However, it is a trade-off with convenience.

10.1.3 Physical Stability

The structure of a camera should be strong and stable enough that the critical internal dimensions of the optical and mechanical subsystems do not change significantly in the course of normal operation and handling. In addition, mechanically fastened joints should not shift or loosen during the life of the camera. Since dimensional precision in some cases goes down to millionths of an inch, these requirements are demanding indeed.

Another mechanical design trade-off exists between making a precision item mechanically fixed versus making it field adjustable. Fixing the item is desirable but it means that the precise setting must be determined in manufacture and then the design must be stable enough that it will never drift out of adjustment in the future. Making it an adjustment means that an error that shows up after manufacture can be corrected but the adjustment itself adds the potential for instability and, of course, adjusting a precise mechanical parameter in the field is a nuisance. The modern approach is to strive for fixing all the precise mechanical parameters at the time of manufacture.

Most cameras use cast metal or (sometimes) plastic structures to meet these requirements. Aluminum castings are generally used in studio cameras, magnesium is used in

Figure 10.3 *Structure and integrated imaging capsule design of the Sony BVP-700/750 cameras. (From Sony.)*

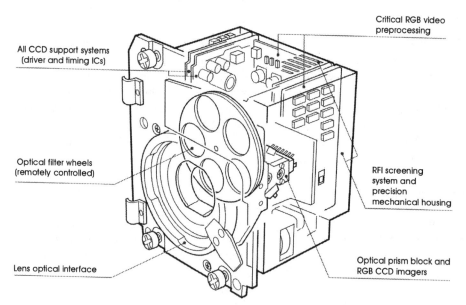

Critical RGB video preprocessing

All CCD support systems (driver and timing ICs)

Optical filter wheels (remotely controlled)

RFI screening system and precision mechanical housing

Lens optical interface

Optical prism block and RGB CCD imagers

Figure 10.4 *Internal details of the Sony BVP 700/750 integrated imaging capsule. (From Sony.)*

high-end portable cameras, and aluminum or plastic is used in home cameras. Still, the philosophy of design must be such to minimize opportunity for errors because even cast structures may have internal stresses or temperature dependence.

An example of high-end studio and portable camera head design is in Figure 10.3, which shows the Sony BVP-700/750 camera series. These units have cast camera structures using aluminum (studio) and magnesium (portable) castings for the basic structure. A unique interchangeable optical assembly called the "integrated imaging capsule" is used. This unit, shown in Figure 10.4, contains the CCDs and their optical system, filter wheels, lens interface, and all electronics that support the CCD scanning and video interface. It allows the same camera to be changed in the field from 4:3 to 16:9 operation; the capsules can be replaced without major electrical or mechanical adjustments being required. This feature also has the advantage that the critical optics and lens interfaces are mechanically separated from the rest of the camera and should be more immune to stresses placed on the camera body itself.

10.1.4 Ruggedness

Cameras are often subject to rough handling in use or transportation. Portable cameras are carried into the field in vehicles or by hand and are subject to vibration and mechanical shock. Even studio cameras get moved about on tripods or dollies, bouncing over cables and other obstructions on the studio floor. They must withstand these abuses without shifting adjustments or sustaining damage.

Ruggedness requires that all fastenings are protected against loosening under vibration or shock. This is especially important for precise adjustments that should not drift in use to the extent that they require excessively frequent readjustment. Further, parts that are normally moved, unscrewed, or otherwise adjusted in normal use should not wear or otherwise become inoperable for the life of the camera.

10.1.5 Operational Flexibility

Cameras are subject to many different uses. Camera packaging should provide for modifications and options that extend the camera's application as widely as possible. In studio cameras, the packaging design should provide for options such as:

- Various types of lenses;
- Choice of viewfinders;
- Different types of operating controls;
- Teleprompter mountings;
- Mounting to various tripods, pedestals, cranes, or dollies;
- Optional mounting means for devices such as monitors, lights, script holders, and so forth.

Similarly, for flexibility in portable or hand-held cameras, they should include features such as:

- Various types of lenses;
- Choice of viewfinders;
- Triax adaptor;
- Weather protection options;
- Carrying handles.

These lists, of course, do not include the electronic options that extend camera flexibility.

10.1.6 Environmental Protection

Studio cameras generally live in a controlled environment when in the studio. However, the same cameras may be used in the field for sporting events or other OB assignments. In those cases, they face the same environmental extremes as portable cameras. Some of these are:

- Operating temperature—in the extreme, this is limited by what the operator can stand. A typical specification is –20 to +45 °C. The camera is expected to be safe and to maintain its performance over this range.

- Storage temperature—this means nonoperating. A typical range is -20 to +60 °C. The camera should withstand extended storage time at these temperatures and be ready to operate correctly when the temperature returns to the operating range.

- Humidity—the biggest problem with high humidity is water condensation within the camera. Camcorders may also have tape handling problems at humidity extremes. Most cameras are specified to operate up from 25 to 85% humidity. At low humidities, there can be problems with static electricity buildup.

- Shock and vibration—these stresses occur primarily during transportation. Most cameras do not specify limits but they are expected to withstand transportation in the back of automobiles or trucks and in various types of aircraft.

- Rain, snow, sleet—most cameras are not fully protected against precipitation. They may stand a brief shower but any prolonged rainfall or snow is likely to penetrate within the camera body and cause problems. In general, if such exposure is expected, the camera should be fitted with a protective cover, which may be as simple as a plastic bag or as fancy as a specially designed weatherproof housing. Whenever covers are added to an operating camera, the effect on the camera's cooling should be considered. Special housings designed by the camera manufacturer will take this into account.

- Electromagnetic interference (EMI)—cameras should be as well shielded agains EMI as possible. This includes both the reception of EMI that may interfere with the picture or sound, and the generation of EMI from camera internal circuits.

- Altitude—this can cause problems with high voltages, especially when the camera has a CRT viewfinder. Altitude also may affect cooling of the camera because the effectiveness of air cooling reduces as the density of air drops with altitude. Generally, cameras will be good up to about 10,000 ft. Going higher than that may call for special care.

Cameras often are taken into extreme environments, particularly when they are used for news service. Generally, they can stand anything the operator can stand. Extreme situations should be approached with special care for both camera and operator.

10.1.7 Thermal Considerations

Heat generation within cameras originates with the IC processing units, the viewfinder (if CRT), and the primary power-conversion unit. The latter converts the input power, whether battery or ac from the camera cable, to the various dc voltages required in the camera circuits.

Most thermal paths in a camera are by convection. The power converter is usually the only unit that may require a conductive thermal path to the camera case. The ultimate heat sink is convection cooling to the air from the outside of the camera case. Within the camera, the issue is to get the heat out to the case. Although some heat buildup is allowable within the camera, some parts perform better if they are kept cool. This applies especially to the CCDs, whose dark current increases with temperature, possibly causing shading and fixed pattern noise problems. Therefore, cooling of the CCDs is a primary thermal design consideration.

Cameras that are used outdoors may be subject to heating by solar radiation. The properties of the case finish may take this into account but, in extreme cases, some form of solar shading may be needed. Cooling may also be a problem if the camera is used in a protective housing, such as a rain sheild or underwater housing. These situations should be handled in the design of the special housing.

10.1.8 Interfaces

Other than the optical interface from the lens to the scene and the operator's interface, a camera connects to its system through a camera cable and it may also have other interfaces to nearby units such as teleprompters, monitors, intercom headsets, or lights. Some of these units (teleprompters or lights) may mount directly to the camera but, in general they all connect electrically through cables. The important consideration is that these interfaces not restrict the movement of the camera and that such movement should not cause undue mechanical stress on cables or their connectors.

These issues are not very significant for a camcorder operating by itself. Other system units in this situation are generally all located on the camera itself or the operator's person and cabling problems are minimal. However, in a studio environment, they may all exist. Therefore, studio camera designs generally have built-in cable stress relief for their camera cable. They also may have devices for routing of other cables or such features may be built into the camera mounting equipment.

10.2 CAMERA DESIGN EXAMPLES

The preceding discussion of design requirements shows the challenges facing a video camera designer. This section presents a number of real camera designs and shows how they deal with some of the requirements.

Figure 10.5 *Panasonic AG-EZ1 hand-held digital camcorder. (From Panasonic.)*

10.2.1 Hand-Held Cameras

Hand-held cameras are generally directed at the consumer market although the high-end units are also used in semi-professional service. Hand-held cameras are occasionally used in news-gathering service on assignments where the convenience of small size and greater mobility may be an advantage.

10.2.1.1 Panasonic AG-EZ1 Digital Camcorder

This camera is shown in Figure 10.5. It uses the DV digital tape format. It can be hand-held as an eyepiece camera, but the eyepiece optics also supports holding the camera away from the eye. The viewfinder unit swivels with respect to the camera body to provide viewing at various camera angles. The camera is held by the right hand around the camera body (a hand strap is on the side of the camera not shown in the figure), and the operating controls are on top of the camera body.

10.2.1.2 Sharp VL-DC1U Digital Camcorder

This camera is shown in Figure 10.6. It is one of Sharp's proprietary Viewcam™ designs using an LCD viewfinder with a swivel mount between it and the camera proper. The operator's right hand holds the camera and operates the lens controls and the VCR record button. The left hand can be used to support the left side of the camera or to rotate the viewfinder portion around a horizontal axis. Thus, the viewfinder can be seen as the operator moves the camera up or down.

This camera weighs less than 2 lb and has all the usual home camcorder features, including image stabilization, automatic focus, iris, and white balancing. The 4-in color LCD viewfinder has 112,320 pixels and has an automatic backlighting feature that provides visibility under any lighting situation. The tape recorder is located in the LCD portion of the camera and is digital, using the DV format; the camera delivers up to 500 TVL resolution off-tape.

Figure 10.6 Sharp VL-DC1U digital camcorder. (From Sharp.)

10.2.2 Shoulder-Held Cameras

Most shoulder-held cameras are camcorders. Some professional units are available as cameras-only; these are intended for integration into studio or OB setups along with studio-type cameras and usually have triax interfaces to the system.

10.2.2.1 Panasonic AJ-D700

The Panasonic AJ-D700 is a professional digital camcorder using the DVCPRO format. It is shown in Figure 10.7. The packaging is mostly in agreement with the typical shoulder-held configuration that was described in Section 8.4.1.1.

Operating controls are on top of the lens where they are convenient to the right hand

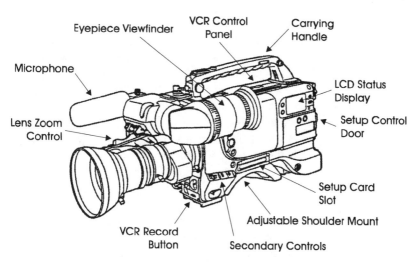

Figure 10.7 Panasonic AJ-D700 digital camcorder. (From Panasonic.)

Figure 10.8 *Sony DNW-90 Betacam SX camcorder. (Courtesy of Sony.)*

that holds the camera. They consist of a rocker switch for zoom control, a VCR record switch, and the manual focus and iris rings on the lens.

Secondary controls are on a panel at the left front of the unit and are intended to be operated by the operator's left hand. They are low enough down that most operators can see them by looking down from the viewfinder eyepiece.

Setup controls are on the left rear of the unit, which requires removing the camera from the shoulder for operation. An LCD panel supports the setup process. Most of the setup controls are behind a hinged door just below the LCD panel. A PCMCIA memory card stores setup data; this is inserted from the left side just above the shoulder support. Obviously, this can only be done when the camera is removed from the shoulder.

Tape cassette loading is at the right rear side of the unit. A pop-out door opens when an eject button on the VCR control panel is pressed; the cassette is inserted from the top. This operation may be done by reaching over the top of the camera with the left hand while the camera remains on the operator's shoulder.

10.2.2.2 Sony Betacam SX

Betacam SX™ is a complete digital news-gathering system by Sony that includes several camcorders, editing recorders, a complete field editing system, and a dockable recorder for use with existing analog cameras. It is based on a Sony-designed digital tape format using the Beta cassette size and making use of 4:2:2 MPEG-2 Studio Profile compression (see Section 13.4.1). It is a very conservative format intended for high performance and

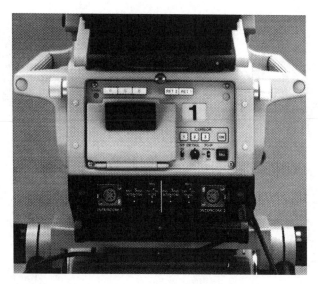

Figure 10.9 *Rear panel of the Sony BVP-700 camera. (Courtesy of Sony.)*

high reliability in field use. Figure 10.8 is a photograph of the DNW-90 one-piece Betacam SX camcorder. This unit also has a conventional shoulder-held camcorder layout.

10.2.3 Studio Cameras

Studio cameras are always mounted on tripods, pedestals, or other mechanical support devices. Thus, their design is not constrained by the weight and size considerations of portable use. This allows larger lenses for higher zoom ranges and higher sensitivity, which is a major reason for the existence of these cameras. Many features are also provided for mounting of optional devices to cameras, such as large viewfinders, extra monitors, teleprompters, and so forth.

Studio-type cameras are also widely used in the field at sporting events where the performance of the large lenses is essential to getting the pictures without disturbing the event.

10.2.3.1 Sony BVP-700

The BVP-700 was shown in Figure 10.1. It is a full-size studio camera that can handle the largest studio lenses, viewfinders, teleprompters, and so on. It uses three 2/3-in CCDs for the highest possible image performance. The camera has integral triax operation and is intended for use with a remote camera control unit where all setup adjustments are performed. Thus, the number of controls at the camera head is reduced compared with self-contained cameras of similar performance.

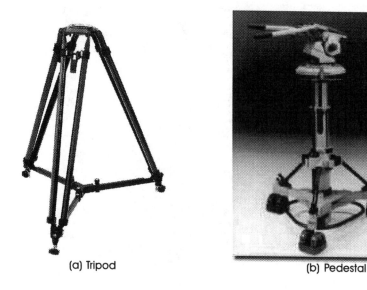

(a) Tripod   (b) Pedestal

Figure 10.10 *Typical camera mounting devices. ((a) From O'Connor, (b) from Vinten.)*

Operating controls of the BVP-700 are on the camera handles, and secondary controls are on a rear-panel area as shown in Figure 10.9. The secondary controls are mostly for choosing viewfinder and intercom modes. There are few setup controls because these are remoted to the CCU.

10.3 CAMERA MOUNTINGS

Studio cameras always require mounting to a support that carries their weight and allows easy movement by the camera operator. Hand-held and shoulder cameras are often supported this way, too, because it improves camera stability and frees the operator to concentrate on other functions. Typical camera mounting devices are shown in Figure 10.10.

10.3.1 Tripods and Pedestals

The most common camera mounting device is the *tripod*, shown in Figure 10.10(a). The typical tripod has separately adjustable legs that allow the tripod to be set up at various heights and to be placed on uneven ground, steps, or hillsides. A pan-and-tilt head at the top of the tripod holds the camera and provides for 360° rotation and a certain amount of up and down tilting. The head is a separate device from the tripod; heads are discussed in the next section.

Additional support is given to a tripod by using a *spreader* between the legs; this may be located anywhere along the legs and it also may be adjustable for the degree of spread. If a spreader is placed at the ends of the legs, it is usually called a *dolly* and may also have

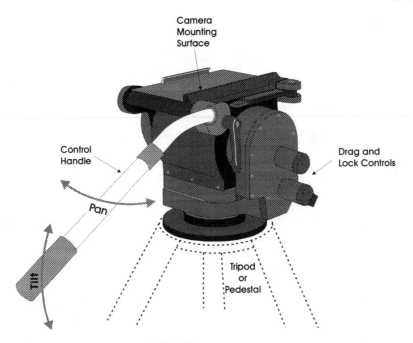

Camera
Mounting
Surface

Control
Handle

Pan

Tilt

Drag and
Lock Controls

Tripod
or
Pedestal

Figure 10.11 *Typical camera pan and tilt head.*

casters or wheels for easy moving of the camera. In this case, the tripod and dolly must be on fairly smooth ground.

A *pedestal*, shown in Figure 10.10(b), is a more capable and convenient device for indoor studio use on smooth floors. It has a rigid, nonadjustable structure that integrates a three-wheel dolly with a single-post mounting of the camera. The camera post is telescoping, and generally contains a hydraulic system that balances the camera's weight. A control ring around the pedestal just below the camera is used to move the camera up or down, simply by the operator lifting or pushing down on it. Because the camera's weight is balanced by the pedestal, it moves smoothly and with little force required. The casters are also steered by the ring; two modes are possible: a *crab* mode, where the three casters are aligned in the same direction and turn together from the control ring; and a *tricycle* mode where two casters remain fixed in the same direction and the third caster is steered by the control ring. These give either straight-line motion in any direction (crab), or rotating motion (tricycle). A foot pedal provides instantaneous switching between steering modes.

10.3.1.1 Pan and Tilt Heads

An additional device is required to control movement of the camera horizontally (panning) or vertically (tilting). This device is called a *pan and tilt head* and is mounted to the top of the tripod or pedestal. An example is shown in Figure 10.11. The head has a quick-

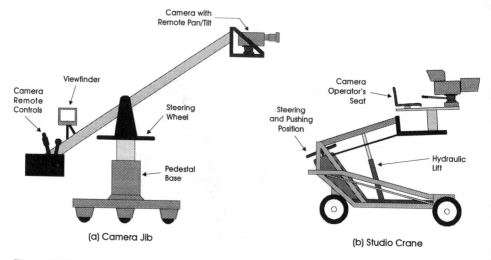

Figure 10.12 *Camera jib and studio crane.*

release mounting surface for the camera and has two-axis movement controlled by the operating handle or handles. The camera weight is balanced by the head when tilting so that it will remain in any tilted position. An adjustable fluid drag mechanism is built in to control the amount of force required to move the camera. The objective is to provide smooth motion as the camera is positioned by the operator.

10.3.2 Camera Cranes and Jibs

More elaborate camera mounting and moving devices are shown in Figure 10.12. Camera *cranes* and jibs provide vertical movement of the camera over a wide range and give the spectacular shots where the camera moves up or down while viewing the same object in the scene. With a studio crane, the camera operator rides with the camera and a second person is required to steer and move the crane. With the jib, the camera is remote-controlled and the operator stands on the ground next to the device with a viewfinder and remote controls for the camera.

10.3.3 Other Mountings

Other mounting devices include the *Steadicam*™, which is a mount worn by the operator that stabilizes the camera against picture jitter caused by the operator's walking, running, or other extreme movements. This was described in Section 5.5.3. Another camera mounting device is the high hat, which is a short tripod-like device for bolting or clamping a pan and tilt head (and camera) to any rigid surface.

Special mountings have been developed for using cameras on helicopters. One device supports the camera outside the aircraft on a remote pan-and-tilt mechanism that is com-

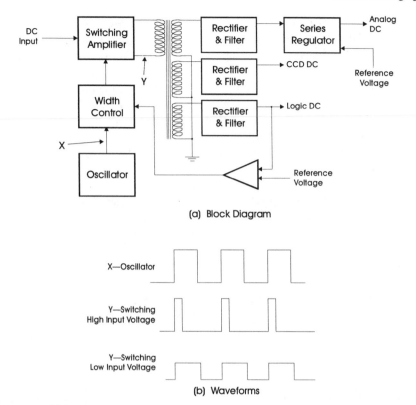

(a) Block Diagram

(b) Waveforms

Figure 10.13 *Block diagram (a) and waveform diagram (b) of a typical power conversion circuit for a battery-operated camcorder.*

puter stabilized against the aircraft motion. Stable pictures at extreme zoom ranges are achieved.

10.4 POWER ISSUES

Although power is not a mechanical issue like the other subjects in this chapter, it has a significant effect on camera packaging, so it is covered here. Power for portable, self-contained cameras comes from batteries, which are discussed in detail below. Other cameras receive their power through camera cables, which were discussed in Section 6.6.

10.4.1 Power Conversion

Source power for the camera, whether from the camera cable or batteries, must be converted to the various dc voltages required by the camera's circuits and electromechanical components. In addition, cameras that provide convenience ac outlets must include an ac inverter for that purpose. A typical power conversion circuit for a battery-operated camera is shown in Figure 10.13. This is an example of solid-state switching-type power conversion, which changes the input power to a high frequency controlled by an oscillator. A switching amplifier switches the input power to a transformer that shifts the voltage to the proper levels for the various output voltages, which are rectified and filtered. Pulse-width control of the converter operation provides voltage regulation against input and load variations. This is illustrated in the waveform diagram, Figure 10.13(b), that shows how pulse-width control compensates for input voltage changes. There is no loss of efficiency at high input voltages because the input current is reduced by shortening the pulse width. The figure also shows how additional series regulation is provided for the more critical analog circuits of the camera.

Even though solid-state power converters are very efficient, they handle all of the power to the camera and are usually the largest source of heat in the camera package. Thus, they should get the prime consideration for heat transfer. Heat sinks for the power converters are usually mounted directly to the outside case of the camera.

10.4.2 Batteries

Portable equipment uses batteries for their power source. The many types of batteries and their correct operation is a major subject, which will only be covered in overview here. The Anton/Bauer web site contains a wealth of information about batteries [1].

10.4.2.1 Types of Batteries

Many rechargeable (secondary) battery technologies are available, ranging from the familiar automotive lead-acid technology through nickel-cadmium (NiCd), nickel-metal-hydride (NiMH), to the various lithium-based technologies. The leading technology for video cameras is NiCd and that is not likely to change for many years. This situation is the result of years of battery-system development and experience with video cameras in professional applications like electronic news-gathering (ENG).

The basic considerations for choosing a battery system are:

- Wattage load and running time—camcorder power drains range from under 10W to over 30W. A camcorder must be in operation all the time that shots are being set up or previewed, even when no tape is running. Thus, battery life must be significantly longer than tape running time. In professional service, a running time of at least 2 hours is desirable. That allows shooting sessions without too much interruption by battery changes. A similar running time seems to be desirable in home application, also. It is generally impractical to charge batteries in the field, both because of the lack of power sources and also because it takes too long. Thus, a field session must carry enough fully charged batteries to cover the longest shooting session. Batteries are usually charged overnight.
- Battery size and weight—obviously, smaller size and weight are desirable if everything else were equal. However, running time usually cannot be traded off for size and weight and there is a limited premium that users will pay for smaller

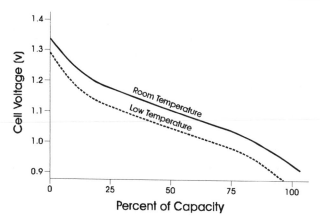

Figure 10.14 *Typical NiCd battery voltage versus life curves.*

batteries. In fact, some users of shoulder-held camcorders think a larger battery is desirable because a heavier battery, which is almost always at the rear of the camcorder, helps balance the weight of the lens on the operator's shoulder.

- Safety and reliability—batteries are chemical devices and are subject to damage, explosion, fire, or release of dangerous substances if improperly applied or operated. Every battery system must give maximum attention to safety considerations to reduce the risk of catastrophe. Modern systems are safe and reliable.

- Charging considerations—the performance and safety of a battery system depends critically on the methods of charging. See Section 10.4.2.3.

- Life-cycle costs—rechargeable batteries do not last forever. There is a limit on the number of charge-discharge cycles before the battery becomes unusable. This can depend on the type of charging system, environmental conditions, and battery handling. Thus, the cost of operating a battery system has many variables. A more expensive initial cost may actually lead to lower operating costs.

These factors are well developed and understood for NiCd battery systems. The one other important technology emerging in the video camera market is NiMH, which provides appropriate capacities with smaller size and weight, although at higher cost.

10.4.2.2 Battery Application

Specifying a NiCd battery for a system involves choosing the number of cells to match the voltage range capability of the system's power converter and then choosing the cell capacity to deliver the desired running time at the system load wattage. Most professional camcorders are designed for input voltages in the range between 11 and 17V. The output voltage behavior of a single NiCd cell is shown in Figure 10.14. It varies during the discharge life and also with temperature. Although the nominal cell voltage is 1.2V, it may go as low as about 0.95V to achieve full capacity at low temperatures. Thus, a conservative battery selection for an 11V minimum is 12 cells in series. At the nominal voltage, this is a 14.4V battery. Fully charged, it delivers more than 16V and, at low temperatures, the voltage could drop to 11V. This uses most of the 11 to 17V range of the typical camcorder power converter.

Home camcorders operate at lower voltages, typically 6 to 9V, but the battery selection criterion is the same—choose a battery whose minimum voltage for full capacity at low temperatures is just above the camcorder's minimum input voltage specification.

Once the number of cells is chosen from voltage considerations, the cell size must be specified to give the desired running time. This is done from the power consumption spec for the camcorder, which is given in watts. A 25-W camcorder that must run for 2 hours, calls for at least a 50 W-hour battery.

10.4.2.3 Battery Charging

By reversing the chemical reaction that occurs in a battery as it is discharged, the battery is recharged. That is accomplished by passing a current through the battery until it is fully charged. Unfortunately, the point of full charge is difficult to detect, especially in a series string of cells where all cells may not be exactly alike. Worse yet, a battery that is charged beyond the full point will convert the charging energy into heat, which can cause damage to the battery, release of gases, or even fire.

Theoretically, the full-charge condition can be detected by monitoring the battery voltage, but that proves unreliable in series strings. A better monitoring method is to measure cell temperatures, preferably for each cell in a string, and the best battery packs have temperature monitoring built in. A computer in the charging equipment checks the cell temperatures during charging and stops charging when cell temperatures begin to rise. Better yet, the computer can monitor individual cell voltage and temperature and generate a charging profile that will bring each cell to full charge as quickly as possible.

The Anton/Bauer InterActive™ charging system contains a computer that reads identification information from a memory in each battery pack and then selects a charging algorithn optimized for that particular type of battery. The algorithms combine several methods of monitoring charging, including time, temperature, and cell voltages. With that feature, this system can handle NiCd and NiMH battery packs on the same charger at the same time.

10.5 AUDIO

Cameras and camcorders generally have some provisions for program audio pickup or intercom. Some of these were discussed in Section 6.8. This section covers the packaging considerations of camera audio features.

10.5.1 Program Audio

Studio cameras have minimal program audio capability because audio is usually handled

Figure 10.15 Wireless microphone receiver mounted to a camcorder. (Courtesy of Sony.)

by the dedicated audio facilities of the studio. However, triax cameras generally provide one or two program audio channels back to the CCU. The camera head contains connectors and microphone preamplifiers so mikes can be plugged into the camera and their signals received at the CCU position. This is applicable when a studio camera is used in OB service where there may be no other audio facilities at the camera location.

On the other hand, portable cameras or camcorders need more complete audio capability because they are usually operating alone in the field. Camcorders generally have means to mount a microphone on the camera as well as a connector for an external microphone to feed the audio recording capability. When using an on-camera mike, it is important that sounds not be picked up from the camcorder itself. That calls for acoustic design considerations on motors, switches, and other controls. This may also be important in studio service so that cameras do not make sounds that are picked up by mikes in the studio. However, it is unusual for an external microphone to be as close to a camera as an on-camera microphone.

There are also provisions in some camcorders to mount a wireless microphone receiver to the camera, as shown in Figure 10.15, which is very convenient in many field situations.

Since there is no other audio unit available, camcorders must also have controls for setting audio levels. An audio level display is also useful; this is often included on the camcorder's status LCD. A further capability is an audio speaker or headphone jack so the operator can hear the program audio. Some shoulder-held camcorders have a small speaker located on their left side at the position where the operator's right ear is located when holding the camcorder.

10.5.2 Intercom

Studio camera intercom facilities were discussed in Section 6.8.2. The principal packing consideration for intercom is the location of connectors for headsets. Two jacks are usually provided, for the camera operator and a *tracker* person. The tracker is anyone who accompanies the camera operator, such as a person pushing a dolly or crane.

Intercom jacks and any other cable connectors on a camera must be located with consideration for the routing of cables from the camera so they do not interfere with normal camera movements. Usually, cables going to external devices should come out at the bottom of the camera near the rear. For intercom cables to the camera operator or tracker, the cables should be no longer than necessary for the normal movement of the people. Other cables going farther away, such as the camera cable, usually should have an anchor point on the camera tripod or pedestal and then run along the floor to their destination. That reduces the stress on the camera itself by pulling on those cables and it reduces the chance of someone tripping over the cable.

10.6 CONCLUSION

Camera packaging is a complex subject encompassing many engineering disciplines. The package designers must deal with the myriad situations in which the camera will operate and provide built-in or optional features to accommodate all the problems camera operators will face. When the package design is done well, the camera is a pleasure for its operator and performs well in all its intended scenarios.

REFERENCE

1.   http://www.antonbauer.com

11

Digital Still-Picture Cameras

The technologies of CCDs, digital processing, digital storage, and digital communication have made possible an electronic still-picture camera with many of the same capabilities as film photographic cameras. The technologies of these cameras are similar to motion-video cameras and, in fact, some cameras contain both capabilities in the same unit. This chapter discusses still-picture cameras, their system, and their uses.

11.1 WHY A DIGITAL STILL-PICTURE CAMERA?

Many still pictures taken today will end up being digitized for inclusion in a digitally processed publication or transmission. It is attractive to shoot such pictures digitally to begin with, thus eliminating film, developing, and scanning. Digital cameras were developed to do this and have other advantages as well.

11.1.1 Description of System

A digital still-picture camera contains a lens that focuses an image of the scene onto a CCD array, followed by ADC, digital processing, digital storage, and digital output means. This is shown in Figure 11.1.

When a picture is taken, the CCD is read out to internal memory in the camera, which may also contain a display for previewing pictures. However, end use requires means for getting the stored pictures out of the camera to a PC or other device for processing, displaying, archiving, transmitting, or printing. That may be accomplished by wire or infrared transmission or by removing a memory module from the camera and placing it in the PC for downloading. Once in the PC, digital stills can be edited, merged with other information such as text, and archived for permament storage. Output from the PC may be in the form of digital files or data streams, which may be sent to other users or printed for hard copy pictures.

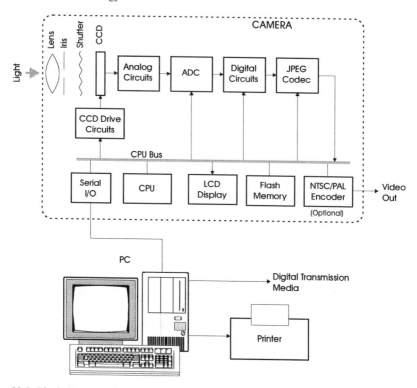

Figure 11.1 Block diagram of a digital still-picture camera and its external system.

The digital still-picture camera is a new product that is still in the early stages of product development. Many manufacturers are designing products and placing them on the market to participate in its growth. Different approaches and features are being tried out but it is too early to tell which will prove to be most important and successful. In many respects, the customers are in a learning stage, too. This is an exciting stage in the development of any new market.

11.1.2 Advantages of Digital Stills

When the intended application for the pictures can accept the limitations, digital still imaging has many advantages.

- Instant results—when a digital still is captured, the picture is available in a few seconds for viewing or other use. There is no developing, printing, and so on.
- Reusable medium—the memory devices used for storing the digital pictures can be erased and reused essentially indefinitely. Nothing is consumed by the process and there is no running cost except batteries.
- Storage capacity—the number of pictures that can be stored in the camera can be very large, depending on the resolution, the degree of compression used, and the

Table 11.1
Picture Storage Numbers for a Typical Digital Camera with 10 MB Memory

Picture Mode	CCD Resolution	Pictures Stored
Uncompressed*	1024 × 768	12
JPEG best	1024 × 768	40
JPEG good	1024 × 768	70
JPEG max. compression	1024 × 768	120
Uncompressed*	512 × 384	48
JPEG good	512 × 384	280

* Based on storing CCD GRGB output format. Others are 24-bit color.

memory capacity in the camera. Some examples are given in Table 11.1.

- Transfer to PC for editing—all digital cameras provide means for transferring pictures to a PC. The capability of PC software for image enhancement and manipulation can then be used to improve and tailor the pictures to suit the end use.

- Lossless copying—digital images can be copied repeatedly with no loss of picture quality. Note that this may not apply when compressed images are decompressed for processing and then recompressed for storage, because lossy compression artifacts may accumulate.

- Low running costs—there is no cost for film or processing. The digital storage media are totally reusable. The only running cost is the cost of batteries for the camera.

These advantages are the reasons for the rapid growth of the market for digital still-picture cameras.

11.1.3 Limitations of Digital Stills

When evaluating a still-picture system, the benchmark is generally 35-mm film. Compared to that system, the limitations or disadvantages of digital stills are:

- Picture quality—except for the very high-priced professional digital cameras, digital pictures have substantially lower resolution than even inexpensive film cameras (see Section 11.2.1.2). Digital cameras also may not deliver the dynamic range of a film camera.

- Light sensitivity—for the same exposure time and picture noise performance, digital cameras are two to four times less sensitive than the fastest films. As in motion video cameras, low-light performance can be achieved with degradation of SNR performance. However, a given amount of noise is more objectionable in still pictures than in motion pictures. That is because the frame repetition in motion imaging helps the eye to visually average out noise.

- Capital cost—entry-level film cameras cost in the tens of dollars. Similar entry-level digital cameras are currently around $200. Digital cameras with performance acceptable for printing at sizes up to 8 × 10 in, cost more like $800. At this writing, the market is highly competitive and prices are dropping rapidly. However, it can be expected that film cameras will always cost somewhat less than digital cameras of comparable quality.

It must be remembered that this comparison is between a mature technology (35-mm film photography) and an emerging technology (digital). Digital still-picture technology is highly competitive among many manufacturers and is rapidly improving in performance and cost-effectiveness.

11.2 TECHNOLOGY

The technologies that are shared between digital still and motion video cameras have already been discussed earlier in this book. The following sections treat these items specifically as they relate to still picture cameras.

11.2.1 Imagers

CCD imagers were the subject of Chapter 3. The fundamentals are the same, but still-picture cameras have several characteristics that can lead to special CCD devices for this service.

11.2.1.1 Scanning of Still Pictures

In a motion video camera, the CCD scans continuously and signal readout occurs concurrently. An important issue in designing CCD architectures for motion video is to make sure the storage of new pictures and the readout process do not interfere with one another. That is not necessary in a still-picture camera and, in fact, the CCD architecture can be simplified by not having concurrent storage and readout. When a still is captured, the CCD is illuminated just long enough to achieve the proper video levels and then the light is shuttered off. Some CCDs have electronic shuttering. The CCD may then be read out from the same registers used for storage. There is no problem of transfer smear because there is no light on the device during transfers. The readout can also occur at whatever rate suits the signal processing and storage capability in the camera; there is no need to synchronize it with anything else.

Still-picture CCDs should use progressive scanning to read all the pixels in a single scan. CCDs designed for this service are progressive scan but some also support interlaced scanning for applications where either still pictures or motion video will be captured by the same camera.

Most digital still-picture cameras use an optical viewfinder, so the CCD need not be active at all until the shutter button is pressed. However, an inactive CCD (even when shuttered from light) will still build up charge in its wells because of dark current. Therefore, immediately before an exposure, the CCD must be scanned once to clear out any dark current stored in the wells. Further scans of the CCD may be used to accomplish white balancing and exposure control. These take only a few milliseconds and do not cause a serious delay in capturing the picture.

11.2.1.2 Resolution

Because of the simplification described above, still-picture CCDs may have only one set of registers and the architectures of FT, IT, or FIT are not always used. As a result, it is economical to build still-picture CCDs at somewhat higher resolutions than motion imagers have. The baseline today for still-camera resolution is 640 × 480 pixels (307,200 pixels). That is acceptable resolution for display on small PC monitors or for printing at sizes up to about 4 × 6 in. Entry-level cameras have fewer pixels, such as 352 × 240, which is acceptable only for Web and Internet applications where the data size of higher resolution pictures would be too large anyway. More and more home and semi-professional digital cameras are offering up to 1,000,000 or more pixels, but of course this raises the price.

11.2.1.3 Achieving Color Performance

Early digital still cameras used a monochrome CCD with a color filter wheel; three sequential exposures were required to take a color picture. This gave excellent results on stationary scenes, but even a small amount of motion causes objectionable color fringing. Current digital cameras all use color CCDs employing the same color filter technologies used in motion video cameras (see Section 3.5).

11.2.1.4 Shuttering

As mentioned earlier, either or both electronic or mechanical shuttering may be used in still cameras. Whether a mechanical shutter is required at all depends on the design of the CCD. CCDs designed for electronic shuttering can disable charge integration by the proper combination of voltages on their electrodes. Because light will remain on the CCD during the readout, these CCDs must have at least an IT architecture to prevent smear problems.

Electronic shuttering has an advantage in that it can easily achieve shorter exposure times than an inexpensive mechanical shutter. Shutter speeds down to 1/10,000 s are possible. Of course, this is usable only in bright light situations.

11.2.1.5 Still-Camera CCD Specifications

Camera CCD design and performance can be deduced in most cases from the camera specifications. A few companies, such as Sony [1], are offering still-picture CCDs on the OEM market. The following items are of interest in still-picture CCD specifications:

- Progressive scan—this is the preferred scanning method for still pictures. All the pixels are read out in a single scan. Cameras that combine motion video capability with still pictures generally have an interlaced-scan readout mode.
- Pixel resolution—the number of horizontal and vertical active pixels. CCDs usually have a few additional pixels around the edges of the sensitive area for black level determination. The total number of pixels (H * V) is also given in specifica-

tions.

- Square pixels—for interfacing with computers, it is preferable to have square pixels. That means the ratio of horizontal to vertical active pixels equals the image aspect ratio.
- Optical size—this is the diagonal measurement of the optical imaging area of the CCD. It determines the optical focal length required for the lens. Most home and semi-professional digital cameras use images sizes of 1/4 or 1/3 in, which keeps the optics small. Professional digital cameras use larger image sizes.
- Architecture—interline-transfer (IT) architecture is most used. As explained in Section 11.2.1.1, a single-register architecture is also possible.
- Shuttering—some CCDs are capable of electronic shuttering. However, most cameras use mechanical shuttering, sometimes in combination with electronic shuttering.
- Readout time—the time to read out all pixels of the image ranges from 1/60 to 1/12 sec. Larger CCDs may take longer to read out. A slow readout time is no problem unless real time motion video performance is required.
- Smear—with electronic shuttering, there may be a problem with transfer smear effects similar to those in motion video CCDs.

Not all of the above CCD specification items will appear in camera specifications.

11.2.2 Digital Processing in Still Cameras

Because of the desire to keep cameras simple to operate and to reduce cost, the built-in digital processing is limited just to those steps required to capture and compress a good picture. Features such as image enhancement, cropping, video effects, and so forth are relegated to a PC. All cameras have means to transfer images to PCs for this purpose. The processes in a typical camera are listed here:

- Analog processing including exposure control, gain control, and white balancing;
- Analog-to-digital conversion;
- Digital processing including highlight compression, gamma correction, aperture correction, and pixel interpolation;
- JPEG compression.

Of course, processing varies in different cameras depending on the particular features they offer.

11.2.3 JPEG Compression

Digital cameras would be much less practical without image compression. All cameras use compression, although some also offer an uncompressed mode as well. In the latter case, the data format is still "compressed" by the color sampling structure of the CCD.

Most cameras use JPEG compression, although they differ somewhat in the features they offer. Because JPEG allows trading between picture quality and degree of compression, most cameras provide two or three choices of how much compression to use. As was shown in Table 11.1, more pictures can be stored when greater compression is used.

11.2.4 Digital Storage

Almost any form of digital mass storage could be used in a digital camera. However, the *flash memory* integrated circuits have become a standard for these cameras. Flash memory is solid-state nonvolatile RAM that has been coming down in cost recently and is seeing application in many products. Most cameras have several MB of flash memory built-in and many provide for the use of removable flash memory cards, too. Memory cards come in the PCMCIA format and also in a newer, smaller, format called a CompactFlash™ (CF) card. A CF card can also be used in PCMCIA slots with an inexpensive adaptor. PCMCIA flash memory cards have capacities up to about 40 MB and CF cards currently can handle storage up to 20 MB. These numbers will surely grow in the future.

Cameras that use the PCMCIA card format may also support hard disks that come in the Type II and Type III PCMCIA format. These are currently available in sizes up to 360 MB and that number will continue to grow. Such capacities allow thousands of still pictures to be stored but, more importantly, they also allow practical amounts of audio or motion video to be stored. Cameras that combine still pictures with audio and video all use hard disk storage.

Theoretically, removable hard disks or flash memory cards can transfer their information to a PC by having a compatible slot in the PC. Although nearly all laptop PCs have PCMCIA slots, very few desktop PCs have them. Maybe this will change with wider use of digital cameras. Since the data storage formats used by digital cameras on memory cards or hard disks are not necessarily the same as a PC's internal storage format, each camera must provide PC software for transfer of its data from a card plugged into a PC.

11.2.5 Digital Communication

Data transfer to a PC is an essential feature of all digital cameras. The high-speed digital data communication market is in an early stage of development and, at present, the only digital interfaces that exist on all PCs are the serial RS-232C format and the parallel printer port. Nearly all digital cameras use RS-232C serial communication even though its fastest data rate of 115,200 bps (on most PCs) is really too slow for this application.

At that data rate, a 1024 × 768 JPEG-best compressed image, which is about 250 kB of data, transfers in about 20 sec. That may not sound too bad until you consider how long it takes to transfer a camera's typical full capacity of 40 such images—13 min. It is even worse when dealing with uncompressed images. The advantage of RS-232C is that it is on all PCs and it is easy and inexpensive to implement in the camera; the disadvantage is slowness.

A few digital cameras provide an interface through the PC parallel port, which can be about 10 times faster than RS-232C. This costs a little more in the camera and the connectors and cables are larger and more expensive than RS-232C. Other camera designers have bit the bullet and designed for even faster transfer methods, such as IEEE 1394 or a PC SCSI interface. In these cases, an adaptor card is necessary to receive the data.

11.2.6 Control

Digital cameras have a lot of features that require controls and complex processes associated with many of them. In addition, they generally have displays to show camera status. This is an ideal application for an embedded microprocessor and all digital cameras have one. The camera CPU typically manages control of the camera and memory transfer operations but it does not participate in the digital image processing or compression. The latter is provided by dedicated digital signal processing chips.

The main image processing in the camera is usually performed by a custom IC designed for that camera. A generic process, such as JPEG compression, can use off-the-shelf chips, several of which are on the OEM market.

A typical architecture of a digital camera with an embedded CPU was shown in Figure 11.1.

11.2.7 Displays

Although most digital cameras use optical viewfinders, they also have electronic color displays so stored pictures can be viewed and erase decisions made. The preferred display is a color LCD, with sizes usually in the range of 1.5 to 2 in. It is difficult to evaluate picture quality on such small displays but a larger display would increase the size of the camera, which is undesirable. For critical evaluation of pictures, it is therefore necessary to transfer the pictures to a PC and view them on a large monitor. With a laptop computer, even this can be done in the field if necessary. The usual problems with viewing LCD displays in bright sunlight or in the dark also apply to this application.

Some digital cameras can operate their LCD display in real time to show the picture before capturing it. This is useful for careful framing, especially in close-up work. However, battery life is shortened by this and the display frame rate may be quite slow.

There is a need on digital cameras to display status information, such as operating modes, amount of space remaining for storage of additional pictures, and so forth. This is usually done with a small monochrome LCD panel having a designed-in custom text and graphics layout.

11.3 FEATURES OF STILL CAMERAS

The heritage of digital cameras embraces both the video and photographic fields. That has provided a rich collection of possible features, which shows up in the diversity of design in the 100 or so cameras already on the market. Because of this, camera comparisons are very difficult. This section discusses the technological aspects of a few of the more important features.

11.3.1 Automatic Focusing

Because of the small CCD optical sizes in digital cameras, it is feasible in low-end cameras to use a fixed-focus lens. However, the degree of close-up capability is limited unless sensitivity is severely traded off by stopping down the lens. Thus, the better cameras all use automatic focusing.

Two approaches are being used: through-the lens AF using the CCD as a sensor, and an infrared rangefinder operating outside the lens. Each of these has its advantages and disadvantages (see Section 5.4).

Some cameras provide a manual focusing mode. That requires the LCD monitor to show the picture before it is captured, which is not available on some cameras. When the LCD is operated this way, it may display a low frame rate because the still-picture readout is slow, and there is always a power concern because this will reduce battery life. Also, focusing with the small LCD picture is difficult at best.

11.3.2 Automatic Exposure

The CCD is generally used for automatic exposure. When the user presses the shutter button, one or more scans are made by the CCD to determine exposure. Finally, a scan is made at the calculated exposure setting and read into memory to capture the shot. Each camera has its own proprietary algorithm for this purpose.

Most cameras do not provide full manual exposure control but they do offer means to adjust the automatic exposure setting by one or more f-stops either way. The Kodak DC120 provides manual exposure control by setting shutter speed only; the iris setting is set to a predetermined value for each shutter speed.

11.3.3 Automatic White Balancing

All cameras have AWB, done by prescanning the CCD between the time of pressing the shutter button and actual capture of the picture. Since there cannot be too many scans during this time because it will cause an excessive delay in capturing the shot, the algorithms are different from the full-feedback approach used in motion video cameras. See Section 5.2.

Camera specifications sometimes include the color temperature range that is accommodated by the AWB system. A typical range is from 2800 to 6500K.

11.3.4 Shutter Speeds

As explained in Section 11.2.1.4, both mechanical and electronic shutters are employed. Mechanical shutter speeds go up to 1/500 or 1/1,000 sec, electronic shuttering can go up to 1/10,000 s. At the other end, some cameras can expose as much as 18 sec under automatic control. That allows taking pictures under extreme low-light conditions, but of course a tripod or other fixed camera mounting is essential.

11.3.5 Automatic Flash

Similar to photographic cameras, many digital cameras have built-in electronic flash. Within the limitations on camera size and battery power, there is usually only enough flash power for scene distances up to 15 ft or so. Some cameras have a *hot shoe* flash connection for interfacing a higher-power photographic flash system to the camera. The hot shoe sends signals to the flash unit to synchronize its operation with the camera's shuttering.

Most flash cameras offer several modes of operation:

- Normal flash—flash usage is based on determination that there is not enough light to take a picture with a hand-held shutter speed (usually greater than 1/60 s.)
- Fill flash—flash operates on every picture, whether needed or not. This is used to fill in shadows in a naturally illuminated scene. The flash intensity is adjusted to be a certain percentage of the natural illumination within, of course, the power limitation of the camera flash unit.
- Off—flash never operates.

Automatic flash adds greatly to the convenience of using the camera for indoor or nighttime picture-taking.

11.3.6 Audio

With the addition of audio input/output facilities, a digital camera can easily record audio into its memory. The most common use of this is to attach an audio clip, say 10 s worth, to each still picture. This can be used to record commentary that is played when the pictures are viewed. Some cameras provide for recording of audio alone, in any length. Considering the limited memory capacity of still-picture cameras (except those that have hard disks,) their audio is usually limited to low quality levels, such as 11 kHz sampling, 8 bits per sample. This level of quality is only good enough for speech recording.

As memory capacity continues to increase, it can be expected that high-end digital cameras will grow into more elaborate audio features, such as stereo recording, higher sampling rates and bits/sample, and sophisticated audio compression algorithms.

11.3.7 Motion Video

Some still-picture cameras are moving across the line that separates them from traditional camcorders by providing motion video recording capability. Even more than audio, the

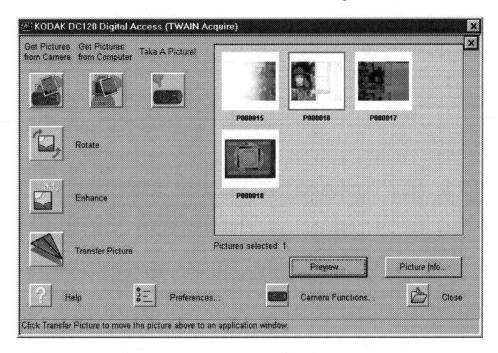

Figure 11.2 *Picture-transfer screen for the Kodak DC120 digital camera. (Courtesy © Eastman Kodak Company.)*

usefulness of this is limited by the storage capacity in the camera. With memory card storage, motion video recording is more of a curiosity than it is useful, but hard disk storage will eventually make this practical. Then, they could be viewed as camcorders that also take still pictures. The Hitachi MPEG camera, described in Section 8.3.2, is an example of this genre. With MPEG-1 compression, it can store 20 to 30 minutes of motion video and audio on a PCMCIA hard disk card. MPEG-1 video is low quality and 20 to 30 minutes is not much, but this does start a trend that will continue as mass storage capacities increase.

11.4 SOFTWARE

Transfer of pictures from the camera to a PC requires software on the PC to accomplish the transfer and to perform image processing on the transferred pictures. The latter is a standard features of many PC software packages that handle images or graphics. This section covers PC software for digital still-picture cameras.

11.4.1 Picture Transfer

Section 11.2.5 described the communication approaches used to transmit pictures from digital cameras to a PC. At present, RS-232C serial transmission is the most popular

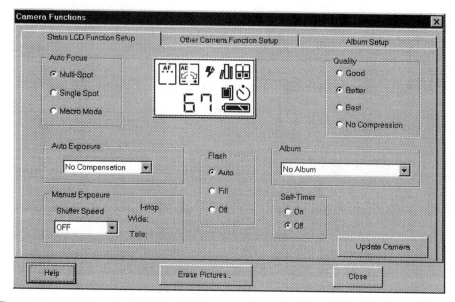

Figure 11.3 Camera setup functions of the DC120 accessed through a PC. (Courtesy © Eastman Kodak Company.)

approach simply because it already exists on every PC. Software bundled with the camera generally provides a separate program for transfer to the PC's hard disk and a *plug-in* module that allows pictures to be transferred within many PC image-processing programs. An example of such a picture-transfer plug-in is shown in Figure 11.2.

The figure shows the main window of the program. The user chooses the "Get Pictures from the Camera" button, which establishes communication with the camera and opens another dialog box (not shown in the figure) listing all the pictures currently in the camera's memory. The user selects which pictures to download and clicks OK in that dialog. Thumbnail images of the selected pictures are then retrieved and displayed in the main window as shown at the right in the figure. The transfer dialog allows some simple processes to be applied during the transfer, as indicated by the "Rotate" and "Enhance" buttons. By selecting a picture from the thumbnails and clicking the "Transfer" button, the user downloads that entire picture into the host program. Once in the host program, the picture can be processed using any of the tools of that program and saved to the hard disk in any format the host program can handle.

The picture-transfer dialog also has a button, "Camera Functions," that accesses the setup menu of the camera and allows adjustment of many camera parameters from the PC screen. This dialog for the DC120 is shown in Figure 11.3. The drawing in the center of the dialog is a replica of the LCD status display on the camera. The number there is the number of pictures remaining at the current compression quality setting. This changes when the Quality selection is changed. Selectors also allow changing the autofocus and flash modes, the shutter speed, and setting the self-timer. Another button, "Erase Pictures," opens a further dialog for erasing one or more pictures from the camera's memory.

Figure 11.4 Screen shot of Adobe Photoshop being used to create a photo montage.

11.4.2 Image Processing

Although the dialog shown in Figure 11.2 offers a few image processing functions, vastly more capability is available by using an image processing program on the PC. These programs range from simple, easy-to-use programs that allow cropping, resampling, rotation, color and transfer-characteristic modifications up to professional-level programs that can create photo montages, and have airbrushing, video effects, graphics, text, and many other functions. One of these is Adobe Photoshop, whose screen is illustrated in Figure 11.4, where it is being used to create a photo montage. Some of the source images are in the windows at the left and the finished montage is shown in the window at the right. Each source image was adjusted for color, brightness, and contrast, cropped with soft edges, and resized for placement into the montage. When finished, the montage is saved in its own file for easy access and use later.

11.5 DESIGN EXAMPLES

This section describes a few current camera designs. The cameras covered here have been chosen for their diversity and they range from low-priced home and semiprofessional

Table 11.2
Comparison of Digital Still-Picture Cameras

Item	Kodak DC120	Ricoh RDC-2	Olympus D-300L	Sony Mavica® FD7	Hitachi MPEG Cam	Kodak DCS460
CCD—Active pixels (H × V)	850 × 984	768 × 576	1024 × 768	640 × 480	704 × 480	3060 × 2036
Image size (in)	¼	¼	¼	¼	¼	1.3
Total pixels	836,400	380,000	786,432	307,200	390,000	6,211,800
Lens focal length	3× Zoom	Wide/normal	Normal	10× Zoom	3× Zoom	Std Nikon
Focusing	Auto	Auto	Auto	Auto	Electronic	Std Nikon
Shutter**	Mechanical	Mechanical	Mechanical	Electronic	Electronic	
Resolution modes	1280 × 960*	768 × 576*	1024 × 768 / 512 × 384	640 × 480	704 × 480 Still / 352 × 250 Video	3060 × 2036
Compression	Kodak	JPEG	JPEG	JPEG	JPEG, MPEG1	None
Mass storage						
Internal RAM	2 MB	2 MB	6 MB	None	None	None
Removable RAM	CF Card	PCMCIA	None	3.5 Floppy disk	None	None
PCMCIA hard disk	No	No	No	No	260 MB	170 MB
Flash modes	Auto/Fill/Off	Auto/Fill/Off	Auto/Fill/Off	Yes	None	Std Nikon
Viewfinder	Optical	Optical	Optical	Uses the LCD	Uses the LCD	SLR
Color display	1.6 in LCD	1.8 in LCD	1.8 in LCD	2.5 in LCD	1.8 in LCD	None
Other displays	Status LCD	Status LCD	Status LCD	None	None	Status LCD
Audio	No	Yes	No	No	Yes	Yes
Sampling rate		8 kHz			MPEG1	
Recording time		86 min max.			4 hr max.	
Motion video	No	1 f/s	No	No	Yes	No
Recording time					20 min	
Computer interface	RS-232C	RS-232C	RS-232C	None	None	SCSI
TV video output	No	Yes	No	No	Yes	No

* Interpolated horizontal.
** Most cameras use CCD shuttering as well as mechanical shuttering.

Figure 11.5 *The Kodak DC120 digital camera. (Courtesy © Eastman Kodak Company.)*

cameras up to full professional high-resolution camera (the Kodak DCS460). Except for the latter, which is priced around $20,000, all the cameras currently cost less than $800 and prices will surely come down as volume builds in the future. The cameras described here and the Hitachi MPEG Cam (it costs about $2,500 and is described in Section 8.3.2) are compared in Table 11.2.

11.5.1 Kodak DC120

The Kodak DC120 [2] is packaged in a flat, horizontal configuration as shown in Figure 11.5. It provides high resolution for its price although the horizontal pixels are interpolated up from the CCD resolution. It includes a good selection of features and has good software for transfer to PCs via RS-232C. Storage is in internal flash RAM and an optional plug-in CF card.

11.5.2 Ricoh RDC-2

The Ricoh RDC-2 [3] is also a flat configuration but it has a flip-up LCD display as shown in Figure 11.6, which makes it look quite different from the other cameras. The resolution is somewhat less than the Kodak DC120 but the Ricoh has audio recording capability. It uses a full-sized PCMCIA memory card. In addition to a RS-232C computer port, it also has a video output for viewing pictures on a TV directly from the camera.

This camera has a "motion" capture mode that runs at 1 frame/s to produce an approximation to motion video. Because of the slow frame rate, the effect is more like time-lapse photography.

Figure 11.6 *The Ricoh RDC-2 digital camera. (Courtesy of Ricoh Corporation.)*

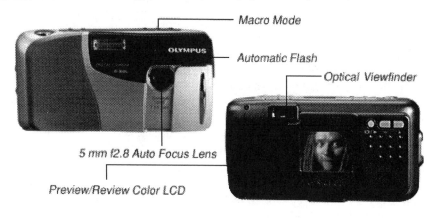

Figure 11.7 *The Olympus D-300L digital camera. (Courtesy of Olympus America, Inc.)*

11.5.3 Olympus D-300L

The Olympus D-300L [4] is configured as a point-and-shoot photographic camera, as shown in Figure 11.7, and it should feel very comfortable to users of those cameras. It provides good resolution and has 6 MB of internal storage. There is no removable storage. Communication with a PC is via RS-232C.

11.5.4 Sony Mavica® FD7

The Sony Mavica [5] cameras are unique in that their storage medium is a 3.5-in, 1.4-MB floppy disk, which is standard on nearly every PC. With JPEG compression, up to 40 images can be stored on a single disk. The FD7 has 640 × 480 resolution and a 10:1 zoom

Figure 11.8 *The Sony Mavica FD7 floppy disk digital camera. (From Sony.)*

lens. The camera configuration, shown in Figure 11.8, is similar to a point-and-shoot photographic camera except that it has to be a little higher to have room inside for the floppy disk.

Since floppy disks are inexpensive, widely available, and readable by virtually every PC, they are in many ways ideal for this application. However, picture transfer from floppy disks is not even as fast as RS-232 and floppy disks, being mechanical, are subject to damage in the field. Regardless, the feature has some attraction.

11.5.5 Kodak DCS460

The Kodak DCS460 [6] is a full professional camera with resolution that comes close to 35 mm film photography. This makes it extremely expensive and is of interest only to those who need (and can pay for) the advantages of digital imaging along with near-full photographic resolution. The camera is based on the Nikon N90 single-lens reflex (SLR) professional camera body, as shown in Figure 11.9. That heritage means that it can make use of all the Nikon lenses, flashes, and other accessories developed for the photographic camera. It also uses the Nikon SLR optical viewfinder that shows the user *exactly* the same image that will be captured when the shutter button is pressed.

With a resolution of 3060 × 2036, a 24-bit uncompressed image is over 18 MB in size. A PCMCIA slot is provided for storage; usually this would be a hard disk. To provide extremely fast transfer to a PC workstation, a *small computer systems interface* (SCSI) interface is provided. SCSI is a multiconductor parallel interface; with it, the camera can transfer an uncompressed image in a few seconds, very much faster than the RS-232C interfaces used in less expensive cameras.

The DC460 has a built-in microphone and telephone-quality audio recording capability that is intended to be used for voice annotation of images.

Figure 11.9 *The Kodak DCS460 professional SLR digital camera. (Courtesy © Eastman Kodak Company, courtesy Nikon, Inc.)*

11.6 CONCLUSION

The digital still-picture camera market is growing rapidly and has already attracted most of the computer accessory manufacturers and the photographic companies. The latter refer to this field as *digital photography*. It is already clear that digital still cameras are ideal for certain applications and they are taking over rapidly in picture-taking for digital end-use applications at low and medium resolutions, such as Web distribution, PC screen display, newspaper photographs, and so forth. Over time, resolution will improve and even more applications will go digital.

REFERENCES

1. http://www.sel.sony.com/semi/ccdarea.html.

2. http://www.kodak.com/daiHome/DCS/DCSGateway.shmtl.

3. http://www.ricohcpg.com/home.html.

4. http://www.olympusamerica.com/digital/index.html.

5. http://www.sel.sony.com/SEL/consumer/mavica/features.html.

6. http://www.kodak.com/daiHome/DCS/DCSIndex.shmtl.

12

Camera Specifications and Measurements

One sign of a mature technology is that all the relevant performance parameters can be specified in hard numbers that are understood uniformly by everyone in the industry. That is generally true for video cameras with the exception that some of the newer digital techniques, such as compression, are still being characterized. This chapter discusses camera specifications and their measurement.

12.1 CAMERA SPECIFICATIONS

The purpose of a specification is to describe a product and its performance with words and numbers. Specifications are essential to the manufacture and sale of products and, after the sale, to the continued maintenance of them.

Specifications may exist at several levels:

- Within a manufacturer's organization, the *design specifications* set forth all the parameters that must be controlled for successful manufacture of the product. Such specifications are the basis for the complete description of the product and all its parts, processes, assemblies, and tests that are contained in the manufacturing drawings of the product. In theory, this set of documents is a complete representation of the product that could be taken to any suitably equipped factory for manufacture. In practice, everything never seems to get set forth on paper and important attributes of the product are contained in the factory's organization, equipment, and experience in building similar products.

- External to the manufacturer, there may be one or two levels of specification. The first level is the specification provided for sales literature use. That is generally a simplified description (compared with the design specifications); it is easily understood and will be useful for customers to evaluate the products of different manufacturers. Thus, industrywide agreement or standardization is vital for such specifications. To a large extent, the items that appear in sales specifications are the attributes on which products compete.

- Some manufacturers provide more detailed specifications and descriptions for their products that are available to serious customers. These documents, sometimes called product information manuals, are more technical and give a complete description of what the manufacturer actually guarantees in the way of product content and performance.

Table 12.1 is a specification at the product information level for a hypothetical camcorder, which is in the format discussed below.

12.1.1 General Specifications

Overall specifications that are not signal performance factors are general specifications. These include items such as size, weight, power sources, environmental limits, and so on. General specifications are also descriptive items, such as the CCD type and pixel counts, optical specifications, lens mounting information, built-in filters and, for camcorders, the VCR type and tape format, and so forth.

12.1.2 Video Performance

Camera video performance is defined by resolution, sensitivity, SNR, smear, shading, and gray scale specifications. These are analog specifications, which are used even with digital cameras. The reason is that analog measurement techniques are well understood and available everywhere, whereas digital measurement techniques are just being developed.

12.1.2.1 Specifying Resolution

Generally, cameras specify limiting resolution values for horizontal and (sometimes) vertical resolution. Professional cameras will usually give a typical value for MTF at one point, generally measured with the camera's image enhancement or aperture correction turned off. The word "typical" in a specification means that this is an approximate specification that is not precisely guaranteed but performance is usually close to that value.

12.1.2.2 Specifying Sensitivity

Sensitivity is specified two ways. The *sensitivity* specification gives the illumination level and lens opening that delivers the specified SNR performance. By convention, this is usually at 2,000 lux, 3,200K illumination with an 89.9% reflectance object set to reference white level. The f-number is then specified; the larger the f-number, the more sensitive the camera.

The second type of sensitivity specification is called *minimum illumination*, which is the lowest illumination level at which a "usable" picture can be achieved. Under this

Table 12.1
Specification for a Hypothetical Camcorder

Item	Specification
General specifications	
Pickup system	3 2/3-in FIT CCD
CCD pixels	1038 × 504 (total)
	980 × 484 (effective)
Optical system	f/1.4 Prism optics
Filter wheels	4-position color filters
	4-position ND filters
Shutter speeds	1/100 to 1/2,000 s
Camera weight (approx.)	15½ lb (with VF, battery, tape, lens))
Dimensions (approx.)	14½ × 10½ × 5¼ in (without lens)
Environmental specifications	
Operating temperature	0 degC to 40 degC (32 degF to 104 degF)
Storage temperature	−20 degC to 60 degC (−4 degF to 140 degF)
Humidity	Less than 85%
Video performance (camera)	
Sensitivity	f/8.0 at 2,000 lux (3,200 degK, 89.9 percent reflectance
Minumum illumination	Approx. 1.9 lux at f/1.4 with +30 dB gain
SNR	62 dB (typical)
H. limiting resolution	900 TVL at center
V. limiting resolution	400 TVL
Horizontal MTF	70 percent at 400 TVL (typical)
Smear level	-140 dB
Registration	0.05 percent (without Lens)
Input/output	
Video out (2 outputs)	1.0 v p-p 75 Ω, 2 outputs
Time code in	1.0 v p-p 75 Ω
Time code out	1.0 v p-p 75 Ω
Genlock in	1.0 v p-p 75 Ω
Audio in (2 channels)	−60 dBu / +4 VU, balanced or high-Z
Viewfinder	
CRT	1.5 in monochrome
Horizontal resolution	600 TVL
VTR specifications	
General specifications	
Tape speed	118.6 mm/s
Recording time	30 min
Audio performance (2 VCR program channels)	
Frequency response	20 Hz to 20 kHz (+0.5 -2 dB)
Dynamic range	More than 80 dB
Distortion	Less than 0.5 percent
Crosstalk (at 1 kHz)	Less than -65 dB
Video Performance	
SNR (luminance)	51 dB minimum
(chrominance)	53 dB minimum

condition, the lens aperture is wide open, the camera gain is increased to maintain full video level, and the video bandwidth may be reduced. SNR is not specified but it is significantly lower that the SNR specification, within the limits of interpretation of what a "usable" picture is.

12.1.2.3 Specifying SNR

SNR is specified under the conditions of the sensitivity specification. Image enhancement and apereture corrections are turned off and the SNR, which is the ratio of the reference white signal to the root-mean-square (rms) noise, is specified in decibels. SNR specifications are also usually "typical."

12.1.2.4 Specifying Smear

Smear is the CCD artifact where extreme highlights in the scene may cause a vertical line to show in the picture above and below the highlight (see Section 3.3). Smear is specified as the ratio between highlight brightness relative to reference white brightness plus the ratio of the smear level in the video to reference white level. It is given in decibels.

12.1.2.5 Specifying Gray Scale

Gray scale is seldom included in camera specifications. Sometimes the gamma correction or its range is specified. Another aspect of gray scale behavior is the handling of highlights (see Section 4.4), but this is also seldom included in specifications. Even though these items may not be specified, they may be measured for the purpose of comparing performance of different cameras.

12.1.2.6 Specifying Shading

White shading is variation of light sensitivity over the area of the picture. Black shading is variation of the signal output when the lens is capped. Shading is sometimes specified; it is usually given as the percent variation of signal output within a specified portion of the picture when the input scene brightness is uniform. It is common to exclude areas near the edges of the picture from shading specifications.

12.1.3 Audio Performance

Cameras or camcorders that have audio systems require the usual specifications for bandwidth, SNR, and distortion. If a recorder is involved, a wow and flutter specification is also necessary. If the camera includes microphone preamplification, the gain and output level ranges should be given.

12.1.4 Input/Output

Camera or camera system (camera plus CCU) specifications generally list all the inputs and outputs of the system and give connector types, signal formats, and so on. This is very useful for judging the capabilities of the camera; if there is no input or output for a particular function, one can be sure that it's not in the camera.

12.1.5 Viewfinder Performance

A viewfinder is the same as a monitor; it is specified by giving its size, type (monochrome or color; CRT or LCD), resolution, and brightness. In addition, any special features such as peaking, zebra, cursors, and so on should be specified. If it is an eyepiece viewfinder, specifications for the viewing optical system may also be given.

12.1.6 Camcorder Specifications

In a camcorder, it may or may not be possible to use the output of the camera without going through the recorder. This is generally possible in professional units but it may not be in consumer-level units. Thus, the video and audio performance specifications for a camcorder may be stated as camera alone or camera plus recorder.

Recorders also have general specifications for cassette tape formats, playing times, rewind or fast forward times, start-up times, and possibly other factors. When the recorder has additional tracks for things such as cue audio, time code, and so on, performance for these items should be specified.

12.2 CAMERA MEASUREMENTS

In the early days of video, camera operation required constant measurement and adjustment to achieve and maintain best performance. Those days are past and today's cameras, even the lowest priced ones, consistently deliver optimal performance with practically no attention at all. Thus, measurement of cameras is now a matter of testing to make purchase decisions or to deal with faulty operation (troubleshooting).

Many camera measurements involve placing test images or charts in front of the camera for viewing through the lens. That may be as simple as pasting a paper chart on the wall or using a rear-illuminated box on which transparency charts are displayed. Measurement is usually accomplished by viewing the camera output on a picture monitor or a *waveform monitor* (WFM), which is a type of oscilloscope designed specifically for video signals (see Figure 12.1). Some measurements are quantitative and hard numbers can be obtained. However, certain camera capabilities are not easily quantified and subjective evaluations must be made. That distinction is covered in the following discussion.

Most measurements require examination of the camera signal in analog form. Very little picture quality testing is possible when the signals are in digital form; they must first

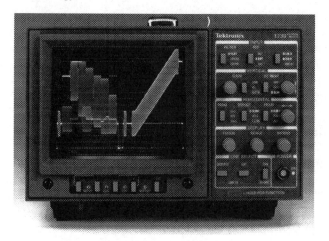

Figure 12.1 *A waveform monitor (Photo courtesy of Tektronix, Inc.)*

be decoded back to analog. The test instruments must synchronize to the scanning of the camera so that pictures can be viewed or specific parts of the waveform can be evaluated.

For highly sophisticated testing, computerized test sets are available that can automatically perform measurements on a signal, statistically correlate results, and present the data on a display or printout. Such equipment is expensive, it is directed mostly at video system and system component testing. However, it could be used for some camera tests, although it would probably not be justifiable for that alone.

12.2.1 Resolution

The two ways of specifying resolution, limiting resolution and MTF, call for different test procedures.

12.2.1.1 Limiting Resolution

Limiting resolution is generally measured by imaging a *resolution wedge* chart, an example of which is shown in Figure 12.2, and viewing the result on a picture monitor. The patterns of converging lines are called *wedges*; they are calibrated for resolution in TVL when the chart is framed by the camera to show full screen. The little arrow marks at the edges of the chart are helpful in achieving the correct framing. Limiting resolution is determined at the points where the lines in the wedges fade out and the value is estimated from the scale at the edge of the wedge. That is shown in Figure 12.3, which is an enlarged view of the horizontal and vertical wedges as viewed by a digital camera. This camera has horizontal limiting resolution of about 450 TVL and a vertical limiting resolution of about 500 TVL.

The chart of Figure 12.2 has four smaller wedge patterns for testing resolution at the corners of the picture. With electron-bean scanned devices such as tube-type imagers or CRTs, resolution may fall off in the corners. However, this is not a problem with CCD

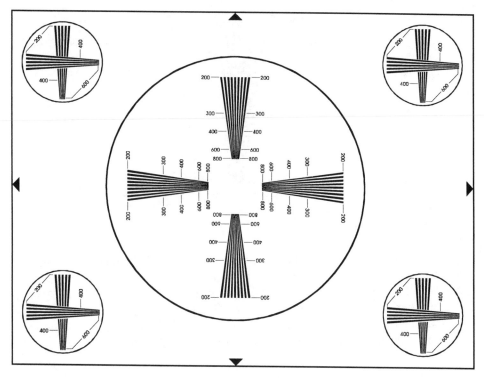

Figure 12.2 *Test chart for measurement of limiting resolution.*

imagers. Lens aberrations (see Section 2.3.2) may cause loss of corner resolution in any type of camera.

When aliasing is present in the system, the wedge pattern will show a sort of moiré

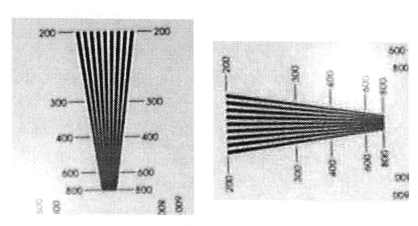

Figure 12.3 *Enlarged views of the horizontal (a) and vertical (b) limiting resolution performance of a digital camera viewing the test chart of Figure 12.2.*

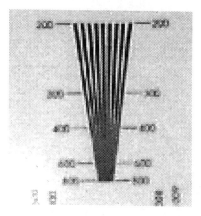

Figure 12.4 *A wedge pattern showing aliasing.*

pattern at the TVL numbers where aliasing is happening. This is shown in Figure 12.4 and it interferes with the measurement so that limiting resolution becomes even more a matter of judgment. Alias interference can be seen beginning at about 250 TVL in the figure, making the wedge not nearly as clear as Figure 12.3(a) nor is the limiting point clearly defined. Generally, the limiting resolution should be taken at the point where aliasing first appears strongly (approximately 300 TVL in Figure 12.4).

12.2.1.2 MTF

In principle, a curve of MTF could be developed from the wedge test chart by examining the video signal with a waveform monitor and measuring the amplitude of the signal from the wedge at various points. However, that is difficult to do and it has a further problem because the wedge pattern is made from sharp-edged lines, so the theoretical output signal from the wedge should be a square wave. Since square waves contain a series of odd harmonics of their fundamental frequency, the amplitudes observed depend on the system's response to the harmonics as well as the fundamental. That causes anomalies in the observed MTF curve.

The correct type of test chart for MTF should have a sine-wave shape to the lines so there is only one frequency represented by each position in the wedge. Better still, instead of wedges, it is more convenient to use a series of bursts of different TVL numbers, making a *multiburst* pattern as shown in Figure 12.5(a). Such a chart shows the horizontal MTF curve directly as the amplitude of the bursts when viewing the signal on a WFM synchronized to the horizontal scan rate. That is shown in Figure 12.5(b).

The first part of the multiburst pattern at the left are white and black bars for setting reference white and black levels. Then, there is a series of bursts of sinewaves at progressively higher TVL numbers. MTF at any TVL number is measured by determining the amplitude of that bar relative to the difference between the levels of the white and black bars.

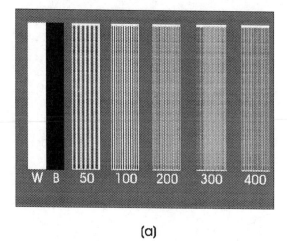

(a)

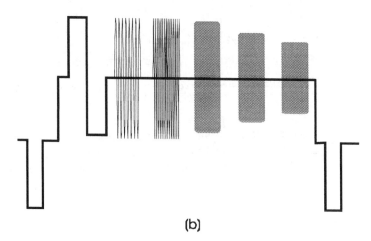

(b)

Figure 12.5 *A multiburst chart (a) and its video waveform (b).*

The multiburst can also be used for measuring vertical MTF by rotating the chart 90 deg. However, aliasing may be more of a problem because of the presence of the scanning lines. Vertical MTF is rarely included in camera specifications but a good camera evaluation should take a look at it anyway. Another type of resolution chart, called a zone plate, allows observation of resolution in any direction [1].

12.2.1.3 Aperture Correction and Image Enhancement

The operation of aperture correction or image enhancement circuits can be observed on the multiburst chart. However, that shows horizontal processing only. With the resolution

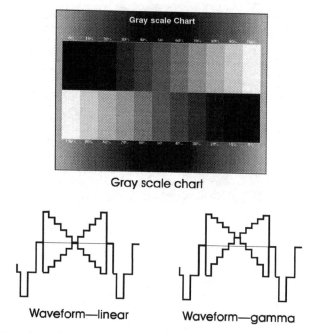

Gray scale chart

Waveform—linear   Waveform—gamma

Figure 12.6 *Testing of gray scale response with a crossed-stairstep test chart. (From [2].)*

wedge chart, the effect of both horizontal and vertical processing can be seen, but of course only a subjective evaluation can be made. The resolution chart also allows evaluation of the symmetry of enhancement—both edges of each wedge line should be equally enhanced.

12.2.2 Gray Scale

Gray scale tests the camera's response to the various levels of brightness existing in the scene. Ideally, the camera should produce a linear response to varying brightness, but modified by the desired (or specified) gamma correction characteristic (see Section 4.8). This is tested by using a *crossed-stairstep chart* that displays steps of brightness, as shown in Figure 12.6.

The stairstep chart contains eleven equally-spaced levels of gray (10 steps) from black to white. Because of the crossed layout of two gray scales, the waveform pattern of the gray scales will cross over at a particular level depending on the gamma of the system. If the system is linear, crossover is at the 50% level. For a system gamma less than unity, the crossover occurs at a higher level as shown in the figure. For a more precise description of the system transfer curve, each step level can be measured on the WFM screen and a curve plotted for the system.

In a color camera, the gray scale characteristics of the three color channels must accurately match to assure proper reproduction of shades of gray in the scene. That is generally not a problem in single-imager cameras but it can be an issue for three-imager cameras. In the case of a composite NTSC or PAL signal, gray scale color match is indicated by the absence of color subcarrier on all the gray levels of the stairstep.

12.2.3 Signal-to-Noise Ratio

SNR is measured by determining the amount of random noise on the video signal in an area of the scene that is uniformly bright. It is often done with the gray scale chart and noise is measured on one or more levels of the gray scale. Because noise may vary on different gray levels, the level for measurement should be specified. The measurement may be done with a special gated noise meter that can be set up to examine a specific area of the scene and determine the noise level there.

It also can be done at less expense using a WFM and evaluating the apparent noise level as a thickening of the waveform trace. The ratio between the observed trace thickness and the actual rms noise level is about six to one, which figure is determined from the statistics of typical noise. This is an approximation, but it is usually reliable within about ±3 dB, which is good enough for most noise evaluations.

If the video waveform display is set up for 100 IRE units* between black and white levels, and N is the number of IRE units of apparent trace thickness, the SNR is calculated as

$$\text{SNR (dB)} = 20 \log (600/N) \tag{12.1}$$

For example, if N is 6, the SNR is 40 dB. This method becomes difficult when the SNR is very high because the trace thickness also includes the width of the scanning spot of the waveform monitor. At 60 dB SNR, N will be 0.6 IRE units, which may be difficult to evaluate. Vertical expansion of the waveform monitor display is helpful in this case.

Noise measurement is affected by the use of aperture correction or image enhancement. These are usually turned off for noise measurement. The measurement may also be affected by the presence of shading or nonuniform illumination, which will cause variation of the signal level within the noise measuring area.

12.2.4 Sensitivity

Because it is always possible to trade sensitivity for SNR by varying the video gain in the camera, sensitivity measurement and SNR measurement are inherently tied together. Thus, sensitivity is determined by establishing the specified illumination conditions and then evaluating the SNR produced by the camera or, (the reverse) by finding the illumination level that produces the specified SNR for the camera.

---

* IRE units are the vertical scale units on video waveform displays. 100 IRE units is the normal range for black-to-white video. The waveform display is calibrated so 100 units represent the actual video voltage level for the system, typically 0.7V.

12.2.4.1 Standard Conditions (Normal Illumination)

A test image is set up to the standard illumination conditions specified in Section 12.1.2.2 (or the standard conditions specified in the camera specification if that is different from the conventional setup). While viewing this image, with manual exposure control and video gain set to normal (0 dB), the lens aperture setting that delivers the specified SNR for the camera is determined.

This requires measurement of the illumination level. With a test chart, the incident light can be measured with an incident light meter, which is placed at the chart location facing the camera and collects the light falling on the chart. If the chart reflectance differs from the value called for in the specification (usually 89.9%), a correction should be made. The other way of measuring the light on the chart is to use a spot light meter that directly reads the reflected light from a highlight area of the chart. A spot meter is like a camera; the operator looks through it at the desired area of the chart and the meter reads the brightness. This method avoids the need for calibrating reflectance.

12.2.4.2 High-Gain Operation (Minimum Illumination)

High gain is used to increase the sensitivity of the camera by trading off SNR performance. Cameras provide anywhere from 18 to 36 dB of extra gain above the normal level, which increases sensitivity by those amounts and reduces SNR by similar factors. Some cameras add additional filtering in the video channel at high gains to improve noise performance at the expense of resolution. This mode is used in news situations or other cases where it is important to capture pictures at almost any cost in terms of picture quality.

High-gain operation is measured by setting up a test chart (usually the gray scale chart) at the specified minimum illumination conditions, increasing the gain, and observing the video level that is achieved. If the camera is meeting its specification, full video level should be achieved. Subjective evaluation of the picture should also be made to judge if the noise performance is "usable" in the sense that the alternative would be to not have a picture at all.

12.2.5 Highlight Handling

As explained in Section 4.4, modern cameras have the ability to compress highlight information that would otherwise extend above reference white level so it is possible to shoot (e.g., an interior room) and show detail simultaneously within the room and through a window to an outside scene illuminated by sunlight. That is done by reducing the slope of the gray scale characteristic near white to bring the highlight details down to reference level or below (see Section 4.4). Because the characteristic below the compressed region is unchanged, the camera can be exposed to show the interior scene correctly.

Although there are seldom specifications about highlight handling, it is useful to test cameras for this characteristic. Cameras may be compared subjectively by setting up the kind of looking-through-a-window scene mentioned above and viewing the results with different cameras. Most cameras have adjustments for highlight compression, so it is a matter of adjusting each camera to do the best job it can when all view the same scene.

For more quantitative evaluation, a test scene may be set up with normal illumination and a normal gray scale range. An illuminated transparency box capable of showing a gray scale transparency at very high brightness is then placed in the scene. The camera is set up and exposed for the test scene without the transparency. Then, the transparency box is turned on and neutral-density filters are placed over it until its highlight brightness comes down to reference white level. Once that is done, any specified highlight level is obtainable by removing the proper amount of ND filters from the transparency box. Thus, the performance of cameras under controlled highlight conditions can be observed.

12.2.6 Picture Geometry

The ability of a camera to correctly reproduce spatially-related objects is its picture geometry performance. Because the spatial layout of the pixels in a CCD is precisely determined during manufacture of the chip, geometrical errors are almost nonexistent in these devices. However, geometry errors still may be caused by aberrations in the camera's optical system and it is sometimes desired to make measurements.

Such measurement is easily done using the test patterns shown in Figure 12.7. The ball chart (on the right in the figure) is placed in front of the camera and the camera's picture is framed to exactly match the chart to the frame size. For observation, an electronically generated grating pattern is mixed with the video from the camera and displayed on a monitor. The camera is positioned critically so the grating lines at the center of the picture fall exactly within the balls. If there are deviations from this condition toward the edges of the picture, the picture size may be adjusted a little by zooming to minimize them. Finally, the picture is examined for any balls where the grating line falls outside the inner circle of the balls, which represents a geometric error of ±1% of picture height. The outer diameter of the balls is ±2% error, grating lines falling entirely outside the balls represent greater error than that.

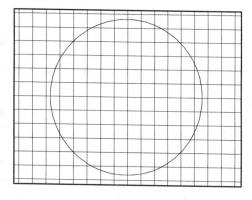 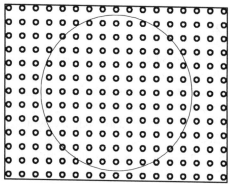

Figure 12.7 *Grating and ball charts for measurement of picture geometry. (From A. C. Luther,* Principles of Digital Audio and Video, *Artech House, Norwood, MA, 1997.)*

This measurement technique can also be used to evaluate picture monitors. In that case, a transparency for the ball chart must be available at the same size as the monitor active screen; it is placed over the monitor screen and the monitor views the electronically generated grating signal. The measurement is by the same method as described above. Observation must be careful to avoid parallax errors.

12.2.7 Registration

The degree to which the three color channels match spatially is *registration*. This is a factor with three-imager cameras; it is usually not a problem with single-imager cameras. Even with three-imager cameras, modern cameras use precision optical assemblies with the devices cemented in place during manufacture. Registration errors less than ±0.05% can be regularly obtained. Registration is observed by imaging a grating chart; one with white lines on a black background is best. Instead of or in addition to the grating, a pattern of white dots on a black background may also be used. Viewing the camera output on a color monitor will show color fringes around the white lines or dots at any places where there is registration error. This test is subjective.

For quantitative measurement of registration, a WFM with dual trace capability and the ability to expand the sweep can be used. The color channels are viewed separately and the WFM is set up to view the relative positions of the color channel signals. By selecting the area of the picture to view, misregistration can be evaluated numerically. This method also works for vertical misregistration.

Registration may be affected by chromatic aberration in the camera lens. That may change with different lens settings of zoom and aperture, so the full range of lens operation should be checked when testing camera registration.

Note that a color monitor may also have misregistration, which cannot be separated from that of the camera. When using a monitor for this purpose, it should first be tested by viewing an electronic grating pattern to assure that it does not have enough error to confuse the measurement of the camera. The WFM method of measurement does not have this problem, but it is much more difficult to perform.

12.2.8 Smear

Smear measurement can be done using a bright light source and neutral-density filters. The light source is set up as described in Section 12.2.5 using ND filters to bring the light level down until the camera output just reaches reference white level. No transparency should be used over the light source. It is convenient to arrange additional scene illumination so the background produces about 20% gray level video; this will ensure that smear effects will not be lost in the blacks.

For the measurement, the ND filters over the light source are progressively removed until a smear effect is seen, as shown in Figure 12.8. The magnitude of the smear signal can be determined using the WFM and smear in decibels may be calculated as

$$\text{Smear (dB)} = 20 \log(100/S * ND) \tag{12.2}$$

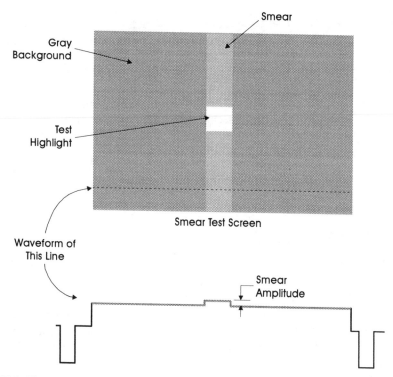

Figure 12.8 *The picture produced by the smear measuring setup described in the text.*

where: *ND* is the attenuation factor of the ND filters that were removed and
            S is the magnitude of the smear signal in IRE units.

It is possible that other highlight overload effects such as blooming or halation may be seen along with the smear or even before the smear shows up. These should be noted in the measurement results.

12.2.9 Shading

Black shading is tested by capping the lens and observing the uniformity of level of the video waveform across the image area using a WFM. Some specifications define one or more rectangles within the total image area for specification. Black shading is caused by dark current in the imager and will become worse at high temperatures. Therefore, the camera should be operated at the maximum temperature in its environmental specification. Many cameras have automatic black shading correction that operates every time the lens is capped. A feedback circuit generates a correction signal that cancels out the black shading. This signal is stored and applied continuously while the camera is shooting. In such cases, the measurement will show the residual shading after the automatic correction has operated.

White shading is tested by focusing the camera on a uniformly illuminated screen. This can be achieved by illuminating a relatively large scene and zooming the camera in on a small part of the scene. The uniformity of illumination can also be checked with a spot light meter. As with black shading, the measurement uses a WFM to examine the video signal amplitude over the specified area of the image. Some cameras have automatic white shading that operates with a stored signal as explained for black shading above. The white shading can only be set up when the camera is viewing a flat white field. The shading measurement, then, would measure the residual after the automatic circuit has operated.

Lens or optical system vignetting may affect white shading. This can vary with different settings for aperture and zoom, so the full range of lens operation should be exercised during a white shading test.

12.2.10 Colorimetry

Colorimetry is seldom specified for cameras except as implied in the statement that the camera meets a particular system specification, such as NTSC or PAL. It is difficult to measure, requiring special test equipment that can generate monochromatic light at all wavelengths in the visible spectrum. It may even be necessary to test outside the visible wavelength range to ensure that response to infrared and ultraviolet is sufficiently attenuated. These measurements are seldom done outside of design and manufacturing laboratories.

Users may compare colorimetric performance of cameras by imaging test charts made up of color bars and examining video waveform amplitudes with a WFM. Since it is difficult to create and maintain color test charts with known chromaticity components, this type of test is useful primarily for comparison of cameras.

An important issue in color matching of cameras in multicamera shooting is the colorimetry of color monitors, since different cameras will be seen on different monitors in the control rooms. Monitors should be carefully calibrated for color temperature and preferably should all be of the same type with the same type and age of CRTs. Actual camera matching, however, should be done with a single monitor, prefereably with a split screen display for showing two or more cameras on the screen at the same time. A color bar test chart can then be used fairly reliably. Some cameras have adjustable masking circuits for colorimetric matching; these should only be operated under such a controlled matching test condition.

12.2.11 Video Compression

The analog video testing that has been described in most of this chapter deals with all the common analog types of picture impairments. However, with the advent of digital video compression, which is used in some cameras today and will be used more in the future, impairments of types never seen in analog systems can occur. Things such as quantization error, blockiness, and motion artifacts are some examples. Although some of these may show up in analog tests, the measurement results may not correlate with the subjective importance of the impairment.

New test techniques are needed for testing the signal performance of compressed video systems. Various companies and industry groups are working on this problem, but so far there are few results. In the meantime, users of compressed video systems must rely on the analog tests combined with subjective evaluations.

In an analog system, video signal performance measurements are made with test charts and test signals and it is assumed the system will behave the same way with signals from real scenes. This assumption is not valid for digital video compression. The signal performance can vary dramatically with scene content, especially between scenes that are simple compared to complex scenes with a lot of detail. Highly detailed scenes may overload a compressed video system and cause the appearance of catastrophic artifacts. Thus, evaluation of compressed video systems, whether by measurement or subjective, should explore the range of expected scene content to locate any places where the system might break down. This is a complex problem that may take a while to sort out.

12.3 CONCLUSION

Camera specifications and measurements have been described in this chapter for the most common parameters. There are many other camera parameters that are not covered, primarily because they are mostly things that are determined during camera design and generally remain consistent between cameras and with camera operation. With the general background of camera technology presented in this book, most readers should be able to figure out ways to perform other tests that may become necessary in their operations.

REFERENCES

1. Luther, A. C., *Principles of Digital Audio and Video*, Artech House, Norwood, MA, 1997, pp. 38–39.

2. Luther, A. C., *Principles of Digital Audio and Video*, Artech House, Norwood, MA, 1997.

13

HDTV Cameras

The subject of high-definition television (HDTV) has been mentioned throughout this book wherever it applied in the technological discussions. This chapter draws all those subjects together and covers HDTV cameras as a whole.

13.1 BACKGROUND

The dream of an improved television system has existed ever since the capabilities of general video technology advanced beyond the original TV standards set in the 1950s and 1960s. Although there are many ways to improve the existing TV systems, the main focus has been to increase the horizontal and vertical resolution by factors of 2 or more; which is called *high-definition television* (HDTV). Other improvements, such as progressive scanning and a wider aspect ratio, more like that in the movies, may also be included.

Serious research on HDTV has been going on for 15 to 20 years and an analog HDTV system has been in limited operation in Japan for several years. However, most systems proposed 5 or more years ago are analog and require substantially more than the 6 MHz bandwidth used by the existing TV standards, which are called *standard-definition TV* (SDTV) systems. Because of the already-crowded frequency spectrum, giving more bandwidth to TV broadcasting is unthinkable.

It took the emergence of digital TV (DTV) to solve the bandwidth problem and, in the last few years, digital HDTV standards have been proposed that use the same 6-MHz channels now allocated for analog TV broadcasting (see Section 4.10.5). A standard has been adopted in the United States but not yet deployed. Another standard, called *Digital Video Broadcasting* (DVB), has been developed in Europe for potentially any form of digital TV signal distribution; it is already in use for satellite broadcasting to the home.

It should be made clear that DTV is not necessarily HDTV and, in fact, that possibility is the basis for some of the controversy currently clouding the start-up of DTV broadcasting in the United States. It is possible that broadcasters might use their DTV channels only for SDTV broadcasting and not implement HDTV at all. This is inconsistent with the goals of DTV and ought not to happen.

13.2 DTV SCANNING STANDARDS

Most of the discussion in this chapter uses by way of example the United States Grand Alliance standards as promulgated by the ATSC and adopted by the FCC. These standards exploit the principle of digital systems that the receiver is capable of storing enough of the data stream that format conversion may be done between the transmitted data stream and the receiver's means of presenting it. Specifically for video, one or more frames are stored in the receiver according to the incoming data stream (this is required to decode the compressed video) and the stored data is read out differently in accordance with the receiver's display scanning capability. For example, the incoming data stream may be interlaced but the receiver, after storing at least a full two-field frame of data, may read that data out to a progressively-scanned display. More elaborate processing allows the receiver to interpolate pixels and lines, so the incoming signal and the receiver's display may actually have different resolutions.

If no bounds were placed on this process, cameras could be scanned in arbitrary formats and receivers would be called upon to convert *anything* so it could be displayed. Obviously, that is too wide open, and the ATSC decided to offer a limited number of scanning formats, covering the range from SDTV to HDTV to motion picture film. These "picture" formats are listed in Table 13.1 and discussed in the rest of this section. However, because of the controversy between the computer industry and the broadcast industry regarding interlaced scanning (see Section 13.2.3), the FCC, in adopting the ATSC proposal, left the matter of picture format choices open to the marketplace. That is not acceptable either, and industry groups have been working to make the choices that will allow reasonable cameras and receivers to be built.

13.2.1 Pixels, Lines, and Frames

The ATSC standard picture formats [1] shown in Table 13.1 support combinations of four pixel and line resolutions, four frame rates, two aspect ratios, and two scanning methods (interlaced and progressive.)

The standard does not support progressive scanning in the 1920 × 1080 pixel format at 60 frames/s. This is a practical limitation based on the data rate capability of the 6-MHz broadcast channel, which is 19.3 Mb/s with the standard modulation. The compression capabilities provided (see Section 13.4.1) allow compression of the lesser formats to data rates much lower than the channel capacity. The transport method (see Section 13.4.2) has the capability of combining data streams for multiple programs up to the channel capacity. Thus, one DTV channel can broadcast one HDTV program or as many as four or five simultaneous SDTV programs.

Table 13.1
Picture Formats Proposed by the ATSC

Active Pixels	Total Active Pixels	Aspect Ratio	Frame Rates*	Prog./Int.
HDTV				
1920 × 1080	2,073,600	16:9	24, 30	P
		16:9	30	I
1280 × 720	921,600	16:9	24, 30, 60	P
SDTV				
704 × 480	337,920	4:3	24, 30, 60	P
		16:9	30	I
640 × 480	307,200	4:3	24, 30, 60	P
		4:3	30	I

*All frame rates also can be NTSC-compatible 1,000/1,001 ratio.

13.2.2 Aspect Ratio

Motion pictures have used a wider aspect ratio for many years. An objective for HDTV was to also provide a wider screen and, after much consideration, the ratio 16:9 (1.77:1) was chosen (see Section 1.6.2). However, the standard also includes SDTV formats that already operate with a 4:3 aspect ratio. Thus, it is a requirement of DTV receivers to accept and properly display both aspect ratios. It is conceivable that some DTV receivers will be manufactured with 4:3 physical display screens because, for some time, 4:3 receivers will be less expensive. A 4:3 receiver would display a 16:9 signal by using black bars at the top and bottom of the screen (a technique known as the *letterbox* format). Similarly, a 16:9 receiver would display 4:3 signals by having black bars at the sides of the screen. However, the attraction of the 16:9 physical screen may cause consumers to wait until they can afford an actual 16:9 receiver. Thus, the market for 4:3 DTV receivers may never develop.

13.2.3 Progressive or Interlaced?

A major expense factor in any video display is the maximum horizontal scanning rate. For progressive scanning, that is equal to the frame rate times the total number of lines. For interlaced scanning, it is the same number divided by two. Table 13.2 shows sampling rates, raw data rates, and horizontal scan rates for some of the picture format choices.

The 2:1 reduction of data and horizontal scan rates offered by interlacing reduces the cost of a display but that comes at a price in performance. Although interlacing is effective for preventing large-area flicker, severe flickering can occur on horizontal or near-horizontal edges of scene objects. This is illustrated in Figure 13.1 for a text object that is 21 pixels high. All pixels in the object that do not have the same information on the lines above and below them will flicker at the frame rate; these pixels are marked with "X" in the figure. This effect, called *interline flicker*, gets worse as the object becomes smaller. With television-type scenes, this effect is usually not too bad because most objects are fuzzy enough that the information on adjacent lines does not differ too much. However,

Table 13.2
Additional Parameters of the Picture Formats

Active Pixels	Total Pixels	Frame Rate, I-P	Sampling Rate (Mhz)	Raw Data Rate* (Mb/s)	H-scan (Hz)
HDTV					
1920 × 1080	2200 × 1125	30-I	74.25	1,500	33,750
		30-P	74.25	1,500	33,750
1280 × 720	1650 × 750	60-P	74.25	1,330	45,000
		24-P	29.70	530	18,000
SDTV					
704 × 480	858 × 525†	30-I	13.50	243	15,750
		60-P	27.00	487	31,500
		60-P	27.00	487	31,500
640 × 480	768 × 525‡	60-P	24.19	442	31,500

* Based on active pixels only, 4:4:4, 8 bits/color.
† These numbers are based on ITU-R BT.601-4.
‡ There is no production standard for 640 × 480 resolution.

with sharply defined computer objects, as in the figure, the effect is unacceptable. Because of this, essentially all computer displays use progressive scanning.

It should be noted that the use of antialiasing on computer-generated objects can improve interline flicker for the same reason that it is improved in "fuzzy" natural scenes. However, this turns the sharp computer text into fuzzy "analog" text, which is a loss of clarity that is not acceptable to computer users.

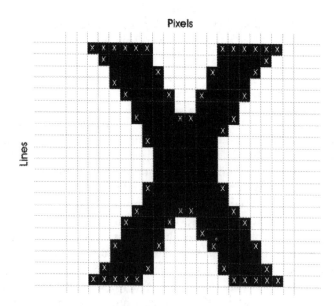

Figure 13.1 *Interlace behavior on a text object. The pixels marked with "X" will flicker.*

A controversy arose in the acceptance of the ATSC standards between the computer industry and the television industry. The former group argues that a new standard should not allow interlacing at all, which would foster the convergence between computers and television. The television industry could not accept that because of the impact of mandated progressive scanning on the cost of low-end DTV receivers. (It is reasonable to expect that the advantages of progressive scanning in receivers can be demonstrated to consumers and those who can afford it will probably buy progressive.) The FCC solved the problem by not solving it—they adopted the standards but did not include standardization of picture formats at all. That leaves that matter to be sorted out by the industry and the marketplace.

13.3 HDTV CAMERA ISSUES

The architecture of an HDTV camera is essentially the same as an SDTV camera except that the numbers for pixel counts and sampling rates are larger. That in itself is enough to make HDTV cameras special (and more expensive), but there are other considerations, such as video compression and compatibility with various DTV picture formats. At this early stage in development, HDTV cameras exist only in the broadcast and production markets and it will be some time before HDTV home cameras will be practical and there will be a market for them. This section discusses some of the special considerations for HDTV or DTV cameras.

13.3.1 DTV Operation

The DTV standards allow operation on a multiplicity of picture formats; DTV receivers are expected to accommodate this and seamlessly display any of the specified formats. Broadcasters may change the format at any time during the broadcast day; especially when using SDTV, they may be broadcasting several programs at once on their channel and the programs may not all be in the same format. This type of operation may also be expected of the cameras in a DTV system—they should be capable of several different formats and should be quickly switchable from one to another almost without interruption of pickup.

Such flexibility is a new feature for cameras and poses a challenge of implementation. For one thing, there is no practical way to switch the number of physical pixels in an imager nor can the aspect ratio be changed at the imager. However, both of these changes can be done electronically in the digital signal path after the imager. Thus, an ideal configuration for a DTV camera would use 16:9 imagers with 1920 × 1080 pixels and electronically convert the output to whatever picture format is desired at a particular time or for a particular program. This is possible today although the hardware size and expense is too great for inclusion in most cameras. That is likely to change as further advances occur in integrated circuit technology. In the meantime, there may be a market for stand-alone format converters as a DTV system component.

13.3.2 Imagers for HDTV

Full-resolution HDTV (1920 × 1080) imagers have been developed by several companies, in 2/3 or 1-in optical sizes. Most of these use interlaced scanning, since progressive scanning at full resolution cannot be transmitted anyway. However, as shown in Section 3.4.4.2, interlacing in a CCD is often accomplished by combining two vertically adjacent pixels from an array that contains pixels for every scanning line. Thus, progressive scanning can be accomplished by changing the off-chip readout circuits. That is advantageous if the signal is to be downconverted to SDTV.

Probably the most important consideration about HDTV imagers is that more pixels in a CCD inherently means lower sensitivity. That is serious because an HDTV imager has as many as five times more pixels than an SDTV imager. The loss of sensitivity may be made up by increasing the video amplifier gain, but SNR is reduced. Therefore, in an HDTV camera, nothing can be traded against sensitivity—the optics must be as fast as possible and the video amplifiers must be of the highest quality. Even so, HDTV cameras will deliver SNR performance 6 to 8 dB poorer than the best STDV cameras.

13.3.3 Lenses for HDTV

The resolution of HDTV imagers is approaching that of many lenses, meaning that the lens contributes significantly to the overall camera MTF. Thus, lenses for HDTV must be of the highest quality. The 16:9 image format also must be considered in the lens design to maintain correction of the lens over the width of the image.

With HDTV, the diffraction limit of resolution in a perfectly corrected lens becomes much more significant. Figure 13.2 shows the theoretical MTF performance for a 2/3-in image format lens. As the lens is stopped down, MTF performance worsens. With an HDTV camera, it is much better (from the resolution standpoint) to reduce exposure by using ND filters than by stopping down the lens.

The depth of field for a lens operating with HDTV is less than the same lens will deliver with SDTV by an amount related to the smaller pixel size (see Section 2.3.1.1). For the same reason, lens aberrations become more visible in HDTV.

13.3.4 HDTV Signal Processing

As already mentioned, HDTV signal processing is similar to SDTV digital cameras except that the sampling rates are higher. Many SDTV cameras operate at sampling rates around 18 MHz, whereas a 1920 × 1080 HDTV camera has a sampling rate of 74.25 MHz. That requires faster IC technology in the signal processing or more parallel processing.

13.3.5 HDTV Camera Cables

An HDTV camera must send the component signals back to the camera control unit through a camera cable. This requires a data rate of approximately 1.5 Gb/s, which can

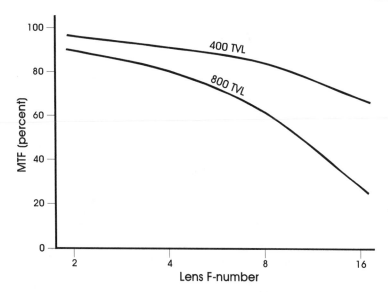

Figure 13.2 *Diffraction-limited MTF performance of a 2/3-in format lens.*

only be achieved by using fiber-optic transmission (see Section 6.6.4.1). Thus, HDTV cameras are one of the proving grounds for high-bit-rate fiber transmission.

13.4 MPEG-2 COMPRESSION

The heart of the ATSC DTV standard is the MPEG-2 video compression and transport system. The basic principles of MPEG were covered in Section 4.10; this section discusses further features of MPEG-2.

MPEG-1 was the first standard developed by the MPEG committee; it was intended for use with progressively scanned images compressed to data rates around 1.5 Mb/s and delivering picture quality that would be competitive with VHS home video. The target application was motion video stored on CD-ROM discs used with computers or dedicated "video-CD" players and the target format was 352 × 288 pixels with progressive scanning at 30 f/s. It was soon found that MPEG-1 was also quite workable with higher data rates and higher-quality pictures and the potential existed for its use in digital broadcasting of conventional television pictures. However, television uses interlaced scanning so there was a need for a version of MPEG that was optimized for such signals. It would also be desirable to add other system-oriented features for increased flexibility for transmission of multiple data streams. Thus, MPEG-2 was born.

MPEG-2 is basically a "toolkit" of compression techniques that are defined by the standardized syntax of a compressed bit stream. Encoding methods to generate such bit streams are not standardized; only the bit stream syntax and decoding methods are standardized.

Table 13.3
MPEG-2 Profiles and Levels

Level	Profiles					
	Simple	Main	4:2:2	SNR	Spatial	High
Low		4:2:0 352 × 288 4 Mb/s I, P, B		4:2:0 352 × 288 4 Mb/s I, P, B		
Main	4:2:0 720 × 576 15 Mb/s I, P	4:2:0 720 × 576 15 Mb/s I, P, B	4:2:2 720 × 576 20 Mb/s I, P, B	4:2:0 720 × 576 15 Mb/s I, P, B		4:2:0, 4:2:2 720 × 576 20 Mb/s I, P, B
High-1440		4:2:0 1440 × 1152 60 Mb/s I, P, B			4:2:0 1440 × 1152 60 Mb/s I, P, B	4:2:0, 4:2:2 1440 × 1152 80 Mb/s I, P, B
High		4:2:0 1920 × 1152 80 Mb/s I, P, B				4:2:0, 4:2:2 1920 × 1152 100 Mb/s I, P, B

13.4.1 MPEG-2 Profiles and Levels

Because of the toolkit approach, many variations of MPEG-2 are possible, covering a wide range of applications. This could lead to a diversity of hardware that might not be interoperable because they selected different combinations from the toolkit. To solve this problem, the MPEG committee introduced the concepts of *profiles* that define certain combinations of techniques that can be guaranteed to be interoperable.

The profiles define subsets of the syntax that a particular decoder is capable of handling. This allows decoders to be built ranging from very simple ones that implement only the simplest parts of MPEG-2 up to a high-profile decoder that is capable of all the features of the standard. Within the syntax allowed by each profile, the concept of level constrains the range of source images that can be processed. This is shown in Table 13.3. Each entry in the table shows the sampling format, the maximum resolution, the maximum bit rate, and the types of frames that are supported.

Profiles and levels are referred to in the form "profile@level," so a system using the Main profile and the Main level is called a MP@ML system. The ATSC DTV standards use MP@ML for SDTV compression and MP@HL for HDTV compression.

The Low level supports quarter-size SDTV pictures, the Main level is for SDTV pictures, and the two High levels are for HDTV pictures. At the Main level, all profiles except 4:2:2 support the 4:2:0 sampling structure. That structure subsamples the color components by 2:1 horizontally and also 2:1 vertically. When used by itself, this approach is satisfactory, but it is undesirable when followed by NTSC or PAL composite analog encoding because it results in a serious loss of chrominance vertical resolution compared to NTSC or PAL alone. That is the reason for the 4:2:2 profile, sometimes called the "studio" profile, which was added for use by broadcasters or others who wanted to use digital compression within their plants but their eventual output format will be analog composite.

The SNR and spatial profiles provide scalability, which means that these bit streams can be decoded at reduced quality by trading off signal-to-noise ratio (SNR scalability) or resolution (spatial scalability). The scalability capability requires the most complex decoders, which is not required in decoders designed for the lesser profiles. In general, a fully qualified decoder will be capable of decoding all profiles and all levels below its designed level (to the left and up in Table 13.3).

13.4.2 The DTV Transport Stream

An MPEG-2 video encoder produces a bit stream containing the bit syntax that represents the compressed picture. This bit stream is parsed into a stream of packets to form what MPEG calls a *packetized elementary stream* (PES). The PES packets are variable-length and a multiplicity of PES bit streams may be generated for video, audio, or supplementary data.

It is the purpose of the transport multiplexer to combine all these bit streams into a single stream for broadcasting. A subset of the MPEG-2 systems multiplexing is used. Fixed-length packets are used in the multiplexing, which allows very flexible utilization of the available transmission data rate. Multiple PESs of any type can be combined within the transport stream. The transport packets are fixed-length (188 bytes); each one contains a header that identifies its type and the particular PES it belongs to. At the receiver, the stream is easily parsed to recover the original PESs for decoding.

Figure 13.3 shows a transport stream built from one video, two audio, and one data elementary streams. It is somewhat exaggerated in that there would likely be many more video packets than shown between each set of audio packets. The multiplexing must contain some buffering to accommodate the rate variations of the input streams. If the total of the input streams is less than the desired rate for the transport stream, empty packets could also be inserted.

The packet headers also contain a sync byte so the transmission demultiplexer can locate the start of packets in the stream and additional bits for a continuity counter so the receiver can tell if a packet has been completely lost in transmission. Another header flag also allows for expansion of the header to include application-specific information.

13.4.3 MPEG-2 in Cameras

Real time MPEG-2 encoding for HDTV is a complex process that has not yet been reduced to custom ICs. Because of the need for a lot of extremely high-speed processing, an encoder is large, expensive, and power-hungry, and is usually a rack-mounted device. There are other reasons why encoding in the camera may not be the right approach because of the problems of dealing with MPEG-2 streams in postproduction (see the next section).

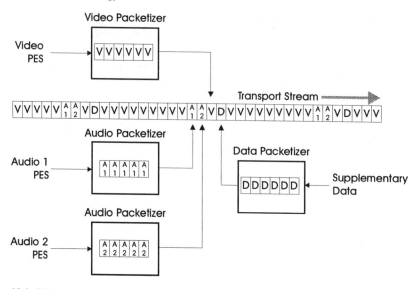

Figure 13.3 *DTV transport packetizing and multiplexing.*

13.5 HDTV Production Systems

As explained in Chapter 6, there is much more to program production than just cameras and this applies just as much to HDTV production. Video captured by cameras must be recorded, edited, and processed for distribution. Signal formats may change along the way in this process, especially so when lossy compression is involved that may cause accumulated degradation of quality when it is applied more than once. HDTV and compression are new to the video production world, and the best methods of handling them are still being developed.

The term "studio" is used here to refer to any HDTV facility, whether it is an actual studio for live production or a postproduction operation.

13.5.1 Signal Formats

Although it would be nice to use the 19.3 Mb/s transport format in the studio, it is unsuited to that because of the artifacts of compression that limit the ability to do any processing on the signal. Switching of the transport stream is also a problem because of the use of interframe compression. Switching can only occur on a I-frame and, even then, consideration must be given to the status of the receiver's data buffer to make sure that switching will not cause buffer overflow at the receiver.

The simple solution is to decompress transport streams for switching or processing and recompress the output signals. This is expensive (especially the recompression part) and leads to accumulated degradation.

On the other hand, using the 1.2 Gb/s uncompressed signals throughout the studio is also impractical. A modest amount of compression is desirable and various approaches are being developed. Ultimately there ought to be a transmission format for HDTV cameras and recorders that runs at somewhere between 100 to 300 Mb/s.

At the same time, research is going on into methods of switching the transport stream so that may be used in situations that only involve switching.

13.5.2 Studio Architectures

Considering the requirement for a limited-compression studio format and the expense of full encoding to the ATSC standards, studio architectures are being designed to minimize the number of encoders and the necessity for transcoding. This leads to a multiformat studio system, which may include HDTV uncompressed signals at 1.5 Gb/s, "studio" compressed HDTV and SDTV signals at 270 Mb/s, and 19-Mb/s ATSC transport streams. This is awkward and it limits the flexibility of the installation, but it seems to be the way things are headed [2].

The other aspect of HDTV systems is that most broadcasters cannot afford to build a new system from scratch for HDTV. HDTV facilities will coexist with present SDTV systems and interconnect with them for DTV broadcasting of SDTV signals. Further, there is need for a system that can "pass through" HDTV signals received from a network. Such a facility, shown in the block diagram of Figure 13.4, may not necessarily originate HDTV signals except for simple things such as inserting the station ID or playing back prerecorded HDTV tapes. However, it should have an encoder so that locally originated SDTV programs, such as news, can be inserted. This figure shows the importance of

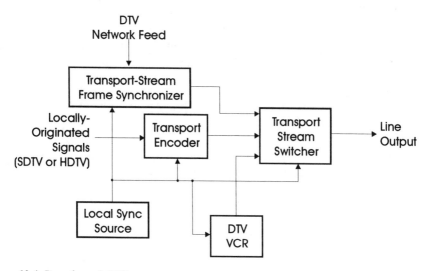

Figure 13.4 *Pass-through DTV system.*

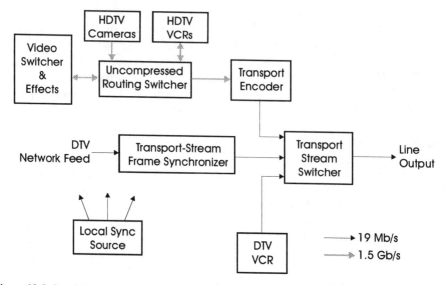

Figure 13.5 *Simplified block diagram of a DTV production-postproduction system.*

having switchers and VTRs for the compressed transport stream. It also shows a transport-stream frame synchronizer for synchronizing the external network feed with the local system.

A full production studio needs to have much more equipment and must switch both uncompressed and compressed signals. Such a system is shown in the simplified block diagram of Figure 13.5. A two-format system is shown, with a production studio and postproduction system running at uncompressed rates. This is connected to a compressed-format system similar to that of Figure 13.4.

A system like this would be less expensive if an intermediate-compressed format was used for the cameras, VCRs, and switcher in the production setup.

13.6 CONCLUSION

It should be evident from this chapter that DTV and HDTV are currently just emerging and techniques and products are still being developed for the first time. Over the next several years, cameras and systems for DTV should become widely available and the expected transition from analog TV to digital TV will be well under way.

REFERENCES

1. The ATSC standards may be downloaded from http://www.atsc.org.

2. Bhatt, B., D. Birks, D. Hermreck, "Digital Television: Making it Work," *IEEE Spectrum*, October 1997.

14

The Future of Video Cameras

The subjects of the previous chapter, DTV and HDTV highlight an important aspect of the future of video cameras—they're going to be digital. The reasons for this were presented in Section 1.5.1, but they bear repeating here because they underlie many of the future trends in cameras, which are the subject of this chapter. Since this chapter is about the future, which has to be speculation or opinion, I am writing it in the first person to highlight the different nature of this discusssion.

14.1 DIGITAL TECHNOLOGY

An all-digital camera ultimately will be higher-performing, more reliable, and lower in cost than an all-analog camera. That is because of the following advantages of digital technology compared to analog technology:

- Once a signal is in digital format, it can be stored, processed, or transmitted with no degradation, if desired.
- Because of the availability of low cost digital memory, processes that require storing lines, fields, or frames of video are practical.
- Complex functions can be designed into integrated circuits (ICs) and produced in large quantities at low cost. An extremely important example of this is the processes for video data compression and decompression, which makes possible recording or transmission of digital video in less bandwidth than required for analog video.
- Once a digital system has been reduced to integrated circuits, it is less expensive, more reliable, and smaller than an equivalent analog system.

Most importantly, digital circuits enjoy the continuing advance of IC technology, which is expected to follow the predictions of *Moore's Law* for at least 10 more years. This "law" is an empirical prediction that was first made in 1965 by Gordon Moore, one of the

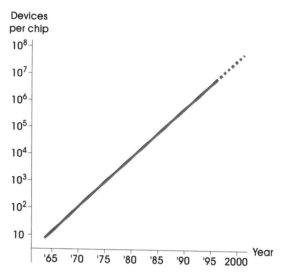

Figure 14.1 *Moore's Law.*

founders of Intel. It states that the number of devices on an IC chip will double every 18 months to two years. This prediction has proved true for over 30 years, as shown by Figure 14.1.

Since the high-volume production cost of an IC chip has not changed very much while the circuit complexity that one chip can hold has increased steadily year after year, the cost per circuit function has dropped dramatically. Most importantly, this applies to digital circuits, which have been able to utilize the extreme complexity that is now possible. This applies much less to analog circuits, which is another argument in favor of going digital.

14.2 TECHNOLOGIES INFLUENCING CAMERAS

As we have seen in the earlier chapters, many technologies are embodied in video cameras. In this section, I discuss the trends in those various technologies.

14.2.1 Integrated Circuits

Modern video cameras, especially portable and hand-held ones, would be impossible without ICs. Figure 14.1 shows the enormous growth in IC capabilities over the last 30 years and projected into the future, but how does that apply to cameras? There are two ways: (1) the capabilities of standard ICs such as memories will continue to increase while their costs will drop; and (2) the number of circuits that can be combined on a single custom-designed chip (an *application-specific IC—ASIC*) will increase.

However, making ASICs more and more complex poses several problems. First, the

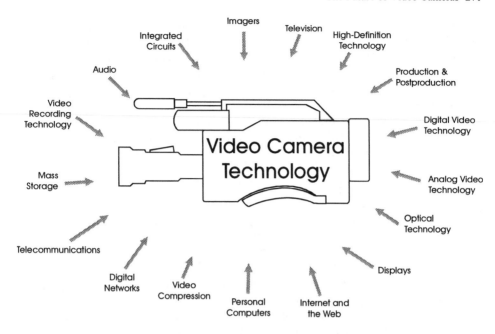

Figure 14.2 *Technologies influencing the future of video cameras.*

design task becomes more difficult and must be supported by better design and simulation systems. Improvement in ASIC design becomes available, but it lags behind the development of the IC capability itself. A second problem is even more important—as ASICs become more complex, they become more specific to a particular application and the manufacturing quantities tend to reduce. That causes their cost to go up because of both amortized design costs and manufacturing per-unit costs—possibly to impractical levels. Thus, there is a design challenge in ASICs to choose architectures that will have multiple applications to keep manufacturing quantities up. So far, the large camera companies have succeeded in overcoming this problem and I think they will continue to do that.

The continued integration of more and more functionality on fewer ICs will provide a platform that can support increased camera capability and features, while keeping costs down. As a result, I think the processing features of both professional and home cameras will become more elaborate and more automatic.

14.2.2 Imagers

To a large extent, imagers (CCDs) benefit from the general advances in IC technology. The ability to fabricate circuits with smaller and smaller physical features allows more pixels to be included on a CCD of a given size. It also allows the architecture of each pixel to be more elaborate, if necessary, to improve performance.

However, CCDs are up against optical and electronic limits that mean they cannot improve in circuit density as much as signal circuits will as the industry progresses up the curve of Moore's Law. Thus, CCDs will not get much smaller than they are today and they will gain only modestly in resolution. But I think what will happen is that they will cost less and this will expand the range of camera applications that are practical. Already we have color cameras for computer use that cost less than $200. This is one of the steps necessary for the eventual wide deployment of video systems such as the video phone. Lower-cost imagers also will contribute to cost reduction and performance improvement of still-image cameras, which will widen that market.

14.2.3 Optical Technology

Optics is a mature field that moves relatively slowly compared with the dynamic field of ICs. However, there is steady improvement of optical materials and techniques. A recent development is the *aspherical* lens, which is the result of a manufacturing process that can produce surface shapes that have fewer aberrations than spherical lenses. This makes possible zoom lenses that are smaller, lighter, and less expensive than conventional units.

Zoom ratios seem to go nowhere except up. The current highest ratio in professional lenses is 70:1; it is not clear how much more than that is really needed, but I'm sure the lens manufacturers will do it if they see a market. However, at the other end, the standard zoom ratio for camcorders, both professional and home, was recently 10:1, but now it has crept up to 15:1. There is probably need for even higher ratios in these markets; as the lens manufacturers can meet the size, weight, and cost limitations, these will appear on the market.

14.2.4 Video Displays

Cameras use displays for viewfinders and they are rapidly moving to LCD displays, at least in the smaller cameras. CRTs will probably prevail for some time in the larger viewfinders because of the brightness issue.

Video displays affect cameras in another way, however. Displays are an important element in promoting the adoption of HDTV by consumers. Large flat-panel displays are being developed for HDTV and, once they become cost-effective, I think they will give a strong impetus to HDTV in the home. That will increase the importance of HDTV cameras, both for broadcasting and for home use. When consumers have HDTV displays in their homes, they will become interested in HDTV camcorders, which will spur development of these devices.

14.2.5 Video Recording Technology

Camcorders require recording technology—today, they all use videotape, mostly analog. Digital videotape formats are available in all markets but, so far, they are important primarily in the professional fields. Even there, several different standards exist, which has kept manufacturing quantities low and prices high.

As professional production moves toward HDTV, video compression in professional recording will become more important to reduce costs and sizes of HDTV recording equipment. However, professionals at present have an aversion to the use of compression because it may introduce picture degradation, particularly when used repeatedly. But the reality is that they will have to overcome this aversion and compression techniques will have to be developed to suit their needs. The market need is there. It is too early to see how the standards situation will turn out.

Another future possibility in recording is the use of disk media in camcorders. Computer hard disks are already being used for video recording for postproduction—in that market, users can afford to buy enough capacity to obtain hours of high-quality recording capacity. The problem is that such equipment is too large, expensive, and power-hungry for camcorder use. On the other hand, the small hard disks that might be physically practical for camcorders can only support marginal picture quality and recording times.

Hard disk recording times, sizes, weights, and costs will continue to improve, but there's a long way to go to meet professional camcorder needs. A more likely cadidate is the proposed recordable DVD (DVD-RAM). This product can record several gigabytes on a single disc, which is not a lot for professional video, but discs can be changed to achieve unlimited capacity, and the discs will be inexpensive. I believe this will be the video recorder of the future for both professional and home use.

14.2.6 Television

In this book, television or TV refers to any system for mass distribution of motion video. Broadcasting (terrestrial and satellite), cable, and videotape are the most important of these systems so far, but other systems are emerging and may be important in the future. These include the Internet and DVD, but others are sure to come forth.

Broadcast and cable TV are beginning a transition to digital transmission (DTV) in many countries around the world. That by itself does not impact cameras greatly except that DTV includes the option for HDTV transmission. As DTV is deployed, the need for HDTV cameras will increase, eventually reaching the home cameras as well as professional ones.

A digital transmission does not necessarily mean digital cameras because all DTV systems will include ADC and encoding capabilities to accept existing analog video sources. Digital cameras will happen mostly because that is a way to improve cameras for any end system, whether analog or digital.

Much speculation is presently devoted to the question of whether the TV and the PC are in conflict for the same markets. Some are saying they will "converge." I discuss this in detail in Section 14.2.10.1.

14.2.7 Production and Postproduction

Production-postproduction is used for essentially all professional program creation and professional cameras are optimized for this style. It is currently not used very much in home video because of the expense and complexity of operation.

The trend in professional postproduction is to perform all operations digitally, storing video on computer hard disks and controlling everything by computer. That provides higher performance and more flexibility than previous analog techniques. It also creates a desire for digital cameras and camcorders with digital interfaces, so that all analog processes can be eliminated outside of the camera.

Home postproduction is becoming more practical as the power of home computers grows to the level that video editing of acceptable quality can be done without a lot of special peripheral hardware. I expect that trend to continue and, although postproduction will never become simple, development of software for home postproduction will address the problem of complex operation. It will become common for home video enthusiasts to produce finished programs from their raw shooting material. As in the professional market and for the same reasons, this will create a demand for home cameras with digital interfaces.

14.2.8 High-Definition Technology

The deployment of DTV and the growth of a market for HDTV equipment has already been mentioned. HDTV displays, processing ICs, and other components will be produced in larger quantities and prices will come down. This will affect other markets that already are using high resolution video systems, such as personal computers (see below).

14.2.9 Video Compression

Digital video, especially in products requiring low cost and small size, depends on compression technology. The MPEG family of standards is rapidly becoming the tool of choice for compression and I expect that will continue for most applications. Dedicated encode and decode chips are already available for MPEG-1 and are under development for MPEG-2. This will reduce the cost and power requirements for using compression in camcorders and stand-alone cameras.

The only real limit on the application of video compression is in the picture quality versus data rate curve. This has a definite relationship for any specific choice of compression techniques. However, since the MPEG standards are toolkits from which a variety of techniques can be selected for different applications, there are many possibilities within MPEG. The other thing about MPEG is that it only standardizes a bit stream and its decoding; encoding can be done by any method that produces a standard bit stream. This leaves many doors open to innovation and I believe considerable improvement is still possible within MPEG.

Video compression is so dynamic a field, it would be folly to say that no standards different from MPEG will be developed—they will. There are always going to be battles for the best compression; as ICs become more powerful, new methods will adapt that power to better video compression. However, MPEG has already attracted such a following that I expect it to be important for many years.

14.2.10 Personal Computers

Digital video has been done in personal computers for more than 10 years by the use of special-purpose add-in hardware. Computers have now reached the point that there is enough power in the main CPU to perform many video tasks in software. Also, the newer CPUs are beginning to have special features built in that provide better support for software digital video. The Intel Pentium MMX feature is an example of this trend. That sort of thing will continue to expand as CPU manufacturers make use of the increased device counts per chip that the progress predicted by Moore's Law puts at their disposal.

With some special hardware, a high-end PC today is capable of storing and playing SDTV-quality motion video and performing nonlinear editing on it. Broadcast stations regularly use PC-based editing systems in news and other local program creation. This trend has nowhere to go except up; system costs will reduce and performance will increase.

By the same token, home PCs will be doing nonlinear editing with hard disk storage and easy-to-use software. Video output may be sent to DVD-RAM for archiving and later playback. All this functionality will only increase the demand for digital home cameras, eventually in HDTV.

14.2.10.1 PC–TV Convergence

As PCs gain high quality video capability, one would naturally ask whether they could replace TV receivers. My opinion is very strong—they will not! I'll explain my reasons below. However, many people do not agree with me, especially those involved in the PC business itself, but I think there are inherent mismatches that will prevent either product from significantly taking over the other's market.

The most important difference between the use of TVs and PCs is in the viewing conditions for displays. A PC, when it is used as a PC, is in a one-on-one situation with its user. A full keyboard is a necessity for serious PC work and the user normally sits close to the monitor with a viewing ratio of 1:1 or so. Contrasting with that, a TV used as a TV for information delivery or entertainment is normally viewed across the room from one or more viewers, at viewing ratios of 3:1 to 7:1. Only a simple remote control unit is needed for TV functions.

A PC may be equipped with a large display that can be placed across the room from the viewer, and a remote-control keyboard provided to sit on the viewer's lap, so that it seems to be operating in the TV viewing situation; but that proves awkward for PC tasks such as word processing, databases, graphics, and so on.

As I am writing this, I am sitting about 16 in away from a 21-in computer display operating at 1600 × 1200 resolution. On that screen, I have a full page display of the page I'm working on, my drawing program for doing the figures, a dictionary program, and so forth. To place that across the room, I would have to have about an 80-in screen to be able to see sufficient detail to work. And, while I am working, which is much of the time, no one else could watch TV in that room.

In the full-room environment required for TV viewing, a PC system would become much more expensive and still, TV viewing and work could not be done simultaneously. This latter problem would cause most consumers to want two products in two different rooms anyway. The most cost-effective way to accomplish that is for one of them to be a TV receiver and the other one to be a PC. Although the electronics of PC and TVs could be converged, the display and use requirements still require two different products.

That is not to say there is no advantages to viewing TV on a PC or viewing PC material on a TV. A student studying and working in his or her room using a PC, may find it very convenient to switch at times to TV-mode and watch specific programs without joining the rest of the family in the TV room. Similarly, there are certain kinds of activities done on a PC that do fit into the viewing environment of TV. The best example is video games, which are often enjoyed by several users at once. This is done today with a game machine connected to a TV receiver. In the future, TVs may have enough PC capabilities to run video games and other multiuser PC applications, although this will be limited by how much the price of a TV receiver can run up and still be sold.

I feel very strongly that TV and PC separate products will continue to sell better than products that combine full-PC and TV functions into one unit.

14.2.11 Digital Mass Storage

The discussion so far should have made clear the importance of digital storage in future video systems. Although digital videotape has been developed and is beginning to be used, disk-based digital storage is much more important to the future. Disk-based storage provides a nonwearing medium (there is no actual physical contact with the medium in writing or reading the data), and any data on the disk can be rapidly accessed. There is no need to go through thousands of feet of tape to find a specific recording on the medium; it can quickly be accessed just by moving heads across the disk surface.

Disk storage is of two main types: magnetic and optical. Magnetic disks are the familiar computer hard disks that have increased so much in capacity over the last five years or so. Today, hard disk capacities up to 8 GB are generally available and these numbers will increase in the future. This capacity can hold up to 15 hours of MPEG-1 video, one hour of MPEG-2 HDTV, or about 5 minutes of 4:2:2 uncompressed SDTV. (These numbers show the importance of compression.)

Storage of 1 hour of uncompressed digital video will require about 100 GB capacity. It is unlikely that single hard disk drives will achieve such capacities in the near future. Hard disk storage of large quantities of high-quality video require arrays of hard disk drives or the use of compression. Until hard disks pass 100 GB per drive, this will be done with digital videotape.

Hard disks have another difficulty—the medium is not removable. The nature of the hard disk drive design requires a sealed enclosure to keep out dirt and dust that might interfere with operation. Hard disks are removable only by removing the entire disk drive. But, many applications require the archiving of large quantities of video—hundreds or even thousands of hours. For video that will be used over many years it is desirable that archiving be of the highest quality, so compression is not preferred. However, I think the numbers indicate that some limited compression will probably always be used; the degree of compression may change over time as storage capacities continue to increase.

The most likely removable storage medium for video in the near future is the DVD and its variations. This optical storage system has a one-sided capacity of 4.7 GB, which can grow to more than 17 GB per disc by use of two layers on two sides. At present, DVDs are created by a mastering process followed by pressing, but soon there will be a recordable version available that can operate from a PC. The proliferation of this product will allow anyone to store high-quality compressed digital video of an hour or more per disc. The capacity will still be too little for storing practical amounts of uncompressed video. That will remain the domain of digital videotape systems for some time yet.

A system like DVD, which is an outgrowth of the compact disc, requires a considerable infrastructure and a large market to support the massive investments in manufacturing necessary to bring prices down. These factors indicate that the market will develop slowly at first and will not mature until manufacturers and customers have made large investments in the same standard. Thus, significant market inertia will mitigate against a lot of changes in standards to keep up with technology developments that could increase storage capacity. This has been the case with CD-ROM, which is now nearly 15 years old. The CD-ROM's capacity of 650 MB now seems small compared to the needs of video, and there shouldn't be much of a problem in the adoption of DVD in computers, at least. This can easily happen over the next 3 to 5 years as computers are naturally replaced to gain the advantages of the advances in CPUs and other computer elements.

What does all this mean to cameras? It is just another driving force toward the continued development of digital cameras.

14.2.12 Telecommunications

As with cameras, telecommunications are going digital. Actually, they are far ahead of the camera market—many digital systems are already in place. The primary telecommunications network in the world is the telephone system; large parts of this are already digital and have been for some time, but most subscriber connections outside of the business world are still analog.

Worse still, those analog connections were designed for voice communication only and have very limited bandwidth. They are used for digital communication through modems, which convert a digital bit stream to a format that fits the analog system's bandwidth and noise performance. Data rates of around 33.6 kHz are the maximum that can be achieved—that number is far too low for any kind of real-time video transfer.

Higher data rates are available through the telephone network to those who can pay for it. Data rates up to 1.5 Mb/s or even higher are used by businesses for internetworking and other purposes. However, the cost of this is out of range for home or small businesses and the existing system would quickly overload if large deployment of high bit rates were attempted.

Because of the growth of the Internet and the World Wide Web (see below), there is a demand for higher-rate digital communication to homes. Expansion of the telephone network could provide that, but it would be an expensive process for both the telephone companies and the subscribers. However, it remains as one of the options.

Other options are being explored to use the cable-TV network, which now connects about 60% of the homes in the U.S., and it important in most other developed countries around the world. Cable-TV has a total bandwidth between 300 and 500 MHz, usually divided into 6-MHz TV channels. Using a modulation scheme such as the U.S. ATSC standard, one 6-MHz cable channel can support data rates up to 38 Mb/s. However, most cable systems are one-way communication, from head end to subscriber and, for the interactivity needed for Internet use, another means must be used for subscriber to head end communication. In most proposals, that medium is the telephone, which may be acceptable because there is much less data transmitted in that direction and the low data rate of telephone connection would not be a problem.

Another option for high-bit-rate Internet connection to homes is satellite communication. Most communication satellites now in use are in *geosynchronous* orbit, which makes them appear stationary from the Earth. However, at the geosynchronous altitude of 22,300 miles, there is about 0.25 sec of transmission delay for a round trip to the satellite and back. This is too much for good interactivity. Several systems are planned to use satellites at lower altitudes. Because such satellites are constantly moving relative to the surface of the Earth, multiple satellites are needed to provide uninterrupted communication. A return communication path through the satellite requires the existence of a transmitting Earth station at the home, which is currently an expensive and tricky device. However, these systems are being designed to support personal communications devices like cellular phones, so one can assume that the transmitting earth station problems will be solved. Thus, satellites may provide higher data rates into the home in 5 to 10 years when some of these proposed systems are in operation.

Telecommunications do not directly affect cameras unless you think of a camera that has an earth station in it and transmits its video and audio up to a satellite. I guess that is a possibility at some point in time, but it still appears to be pretty far out. Otherwise, the growth of high-bit-rate worldwide communication will increase the demand for video and for video devices such as cameras—and they will have to be digital.

14.2.13 The Internet and World Wide Web

Motion video distribution via the Internet is currently practical only among users that have high data rate (above 1 Mb/s) connections to the Internet, or with internal company networks that use Internet protocols—called *intranets*. Even there, it requires massive video compression and relatively low-resolution pictures (e.g., as provided by MPEG-1).

Few cameras are likely to be built for that application alone—most cameras will strive for higher performance to increase their range of application, so the extra compression would usually be provided by a box external to the camera.

Telephone connection to the Internet via modems, such as now exists in many homes and small businesses, is not practical for transmission of video of quality acceptable for general use. Some equipment is available for video conferencing over standard telephone circuits, but the picture quality is very poor, with low resolution and frame rates not much higher than 1 frame/s. These systems often include dedicated low-resolution cameras, but that is not an important camera market simply because of the low picture quality. Higher-quality video conferencing systems require higher-bit-rate connections that are still too expensive for home use.

The World Wide Web protocols allow integration of audio, video, animation, text, and graphics into single entities that can be accessed interactively. The practicality of this is at the mercy of the telecommunications limitations discussed above, which generally exclude motion video. However, the other signal forms are practical, including still video images, if there is not too much resolution. Massive programming energy is now going into the creation of Web pages including these features. This is an important market for still-image cameras, even at low resolutions. That will continue to sell cameras and, as communications capabilities improve, resolution of still images will be increased and eventually motion video will also be practical.

14.3 FUTURE CAMERAS

Based on the foregoing discussion of technologies and markets, certain predictions can be made about future cameras and their markets:

- Future cameras will soon all be digital and have digital interfaces. The deployment of DTV and other digital communication channels guarantee this.
- As DTV proliferates, HDTV cameras will become more important. Because of the existence of SDTV options in the DTV standards, HDTV cameras will also have the feature of downconverted SDTV outputs.
- Cameras in general, for equivalent capability, will become smaller, more efficient, and less costly.
- Home camcorders will first become digital, and later they will have HDTV capability.
- The DVD-RAM recording medium will be important to both professional and home video. It will be included in future camcorders.
- The video camera will be an ubiquitous device.

The last statement deserves some further explanation. All the improvements and cost reduction of video cameras will significantly broaden markets for cameras. Beyond the fields of professional and home program creation, video cameras will be dedicated to video conferencing, video telephones, surveillance and security, medical imaging, personal communication, hobbies, and more.

14.4 CONCLUSION

Video cameras are already developed to a high level of performance and are reliable and inexpensive devices. But this is only the beginning. New systems, new techniques, and new packages will ensure a future of camera advances that will raise them to new heights and new applications.

Glossary

3:2 pulldown In a *telecine system*, a type of film motion that converts 24 frames/s film to 30 frames/s video.

4:2:0 In a *digital component* system, where *color-difference* components are subsampled 2:1 both vertically and horizontally.

4:2:2 In a *digital component* system, where *color-difference* components are subsampled 2:1 horizontally only.

4:4:4 In a *digital component* system, where there are no subsampled components. This format can be RGB or YC_RC_B.

aberrations Image impairments caused by the lens. See *spherical aberration, coma, astigmatism, curvature of field, pincushion, barrel, chromatic aberration.*

active scan interval In a *scanning* system, the group of *pixels* that represents useful video information. The rest of the pixels are usually black and are part of the *blanking intervals.*

acuity The ability of the eye to distinguish find detail. On the average, objects spaced about 1 minute of arc can be distinguished.

ADC See *analog-to-digital conversion.*

additive color A color reproduction system that reproduces by combining color light sources of the *primary colors.* See also *subtractive color.*

Advanced Television Systems Committee (ATSC) An organization in the United States having the objective of standardizing advanced television systems, including the new *digital television* system adopted by the Federal Communications Commission.

algorithm The definition of methods used in a specific process, such as an algorithm for video compression.

aliasing In a sampled system, distortion caused by sampling input frequencies above the *Nyquist limit.*

analog A system that represents signal quantities on a continuous scale.

analog-to-digital conversion (ADC) The process of converting an analog signal to a digital representation. It includes the steps of *sampling, quantizing,* and *encoding.*

aperture correction The process of correcting for *modulation transfer function* loss caused by the *sampling* process.

application-specific IC (ASIC) An integrated circuit (IC) designed for a specific task, such as a camera signal processing IC.

ASIC See *application-specific IC.*

aspect ratio The ratio of the width to the height of a picture.

aspherical lenses Lenses made with a special nonspherical surface shape to reduce *aberrations.*

astigmatism A lens *aberration* where horizontal and vertical object content do not come into focus at the same time.

asymmetrical compression A *compression* system where different amounts of processing are required for *encoding* and *decoding.* Many systems have a complex encoding process but their decoding process is simplified to reduce cost at the receiving device.

ATSC See *Advanced Television Systems Committee.*

attenuation In a camera cable, the loss of signal energy in going through the cable. Loss is frequency dependent, being higher at higher frequencies and, for a given frequency, it increases directly with cable length.

automatic exposure In a *camera,* a system that automatically adjusts exposure according to a predetermined *algorithm.*

automatic focus In a *camera,* a system that automatically adjusts focus to obtain optimal focus in a predetermined region of the picture.

ball chart A test chart used in conjunction with a grating chart to measure picture geometric distortion.

bandwidth The amount of frequency spectrum required by a system. Bandwidth is often associated with a *baseband system,* where it equates to the highest frequency component contained in the signal.

barrel distortion A lens *aberration* wherein the magnification reduces away from the center of the image. It appears as a distortion where the corners of the picture are pulled in.

baseband system A system whose *bandwidth* begins at a low (e.g., less than 60 Hz or even zero frequency).

bidirectionally predicted frames In a *video compression* system, frame reproduction based on both the previous and the next frame of the video sequence. Bidirectional frames (B-frames) provide the most compression. See *predicted frames* and *intracoded frames.*

binary In a *digital* system where the *symbols* have only two values—one or zero. A binary symbol is called a *bit.*

bit See *binary*.

bits per sample In a *sampling* system, the number of *bits* used to *encode* each sample. Abbreviated "bps."

black burst signal In a *composite video* system, a signal that contains sync pulses, color burst, blanking, but no picture signal (black).

black level In a video system, the signal level or value representing the darkest (black) areas of the picture.

blanking interval In a video system, the portions of time in the signal allowed for horizontal and vertical *flyback* of the display. *Analog* video systems always have blanking intervals, but *digital video* systems may not have blanking intervals.

brightness To a human observer, is the sensation of intensity in a picture. See *luminance* and *saturation*.

camcorder A unit that combines a *camera* and a video recorder in one package. Camcorders are generally meant to be portable and carried or held by one person.

camera A unit that views a scene and captures a picture of it. Cameras may be based on photographic film, or they may be electronic. They may capture stills or motion sequences. Photographic cameras inherently store their pictures on the film, but electronic cameras may deliver a signal into a system and/or store pictures on an electronic recording device.

camera cable In a two-piece camera system, connects the *camera head* to the *camera control unit*. See *fiber-optic cable, triax cable*.

camera control unit (CCU) In a two-piece *camera* system, the unit that interfaces the camera to an external video system. It also provides remote control of *camera head* functions.

camera head The pickup unit of a two-piece *camera* system. See *camera control unit*.

candela A unit of *luminous intensity*. A light source emitting energy uniformly in all directions with an intensity of 1 candela emits 1 *lumen* per steradian (unit solid angle.) One candela is defined as the luminous intensity of $1/600,000$ m^2 of projected area of a blackbody radiator operating at the temperature of solidification of platinum and a pressure of $101,325$ n/m^2.

cathode-ray tube (CRT) A display device using a phosphor coating scanned by a high-velocity electron beam in a vacuum.

CCD See *charge-coupled device*.

CCU See *camera control unit*.

charge-coupled device (CCD) The solid-state *imager* device used in nearly all modern cameras. Light from the scene is focused on the active area of the CCD where it is converted to electric charge, which is accumulated in each *pixel* cell of the device. Scanning readout of a CCD is accomplished by moving the charge image so formed both vertically and horizontally to sequentially deliver the pixel charges to an output cell.

chromatic aberration An *aberration* where the *focal length* of a lens varies with *wavelength*, resulting in images of different sizes for each of the primary colors.

chrominance In a *composite video* system, the signal that transmits the *color-difference* components.

CIE An acronym for *Commission Internationale de l'Eclairage*, which is an international standardizing body in the field of illumination.

CIE chromaticity diagram A diagram developed by the *CIE* that shows colors in terms of two parameters X and Y. The locus of spectral colors has a horseshoe shape on this diagram. All other realizable colors have points falling within the horseshoe curve.

circle of confusion In an optical image, the diameter of the focused image from a point of light in the scene.

circular polarization In an electromagnetic wave (radio wave, heat wave, light wave, and so on depending on frequency), causes the electric field vector to rotate in a plane perpendicular to the direction of propagation.

clamping In analog video systems, a process of setting some portion of the signal to a fixed level to restore the DC component. Usually a part of the horizontal blanking interval is devoted to clamping.

coaxial cable A two-conductor cable that has a central conductor surrounded by a uniform-thickness layer of insulation, which is covered by a coaxial sheath that forms the second conductor. The central conductor is used for the signal and the coaxial sheath forms the return or ground conductor.

color correction In a video system, the process of adjusting the color reproduction by linear or nonlinear *masking* between the *component video* signals.

color temperature The naming of a near-white color as the temperature of the blackbody radiator having the closest color. For example, incandescent illumination is approximately 2850K and sunlight is approximately 6500K.

color-difference signals In a *component* color video system, the two signals that represent the color and color intensity of the scene. Color-difference signals go to zero for white regions of the scene. See YC_RC_B *format*.

color-under A system used in analog home video recorders where the *chrominance* information is recorded in the same *track* but separate from the *luminance* information by placing it in a frequency range below the spectrum of the frequency-modulated luminance information. This method allows a very inexpensive means for time-base correcting the color information but not the luminance information.

colorimetry The science of color. In video systems, the word often refers to the determination of the color properties of the system.

colorist A person who operates *color correction* equipment, usually in *telecine* systems.

coma An *aberration* where off-axis point objects focus to points stretched in the radial direction from the center of the image. It is caused by different *focal lengths* for light rays passing through the lens off-axis.

complementary colors In a color reproduction system, colors produced by combining the *primary colors* two at a time. For example, in an *RGB* system, the complementary colors are magenta (red-blue), cyan (blue-green), and yellow (red-green).

component video A video system that separately processes and transmits the three primary color channels. See *RGB* and *YC_RC_B format*.

composite video A video system where the three primary color channels have been combined in some way to send them through a single transmission channel. See *NTSC*, *PAL*, or *SECAM*.

compression In digital systems, any process that reduces the amount of data needed to transmit information. See *lossless compression* and *lossy compression*.

continuous motion transport In a *telecine* system, a film transport that moves the film smoothly at a continuous rate of speed. See *intermittent transport*.

correlated double sampling In a *charge-coupled device* camera, a method for extracting the output video signal while rejecting the noise caused by the need to reset the output gate for each pixel.

CRT See *cathode-ray tube*.

curvature of field A lens *aberration* where off-axis objects focus to different image planes along the axis depending on the distence off-axis.

DAC See *digital-to-analog conversion*.

data rate In a digital system, the transmission rate of bits, usually given in bits/s, kilobits/s (kb/s), or megabits/s (Mb/s).

DCT See *discrete cosine transform*.

decoding In an encoded system, the process of undoing the *encoding* and recovering the reproduced signal.

depletion region In a *charge-coupled device* cell, a region that has an affinity for free electrons. It is the region where the free electrons produced in the cell by photoelectric action are collected until scanned by the read-out mechanism of the device. The depletion region may also be called a *potential well*.

depth of field In a lens system, the range of object distances that appear in focus in the image at a fixed focus setting. With a *pixel*-based imager, depth of field is the range of object distances where the *circle of confusion* is smaller than the pixel dimensions.

dichroic mirrors A color light-splitting means where thinly coated mirrors are designed to selectively reflect or refract light depending on its *wavelength*.

diffraction limit In a lens, the limit of resolution caused by diffraction at the *iris* of the lens. In general, lens resolution increases with larger iris openings.

digital A system where signals are represented by *symbols* having a discrete scale of values. For example, symbols in a decimal digital system have 10 possible values or in a *binary* digital system they have only 2 values.

digital component video A *digital video* system where the video components are separately *digitized*. See *component video*.

digital filter In a *sampled* digital system, a process that manipulates the sample values so as to modify the frequency domain of the information content of the samples. Typically this is done by accumulating samples that have been delayed and weighted by different values. See *finite impulse response filter*.

digital television (DTV) A television system where the transmitted signal is digitally encoded. See *encoding*.

Digital Versatile Disc (DVD) A new digital optical disc format for prerecorded audio, video, and computer use. It is the next generation of the compact disc. DVD-ROM and DVD-RAM (recordable) versions will also be available.

digital video Any video system where the signals are digitally encoded.

Digital Video Broadcasting (DVB) A standard for digital television developed in Europe. The DVB standard is used worldwide for satellite broadcasting of television.

Digital Video Cassette (DVC) A digital videotape system developed for home camcorder use. It uses a small cassette holding ¼-in tape.

digital-to-analog conversion (DAC) The process of converting a digital signal back to analog representation.

digitize A synonym for *analog-to-digital conversion*.

digitized composite video The *digitizing* of analog *composite* formats such as *NTSC* or *PAL*. This is often advantageous when digital equipment is used in an otherwise analog video system.

discrete cosine transform (DCT) In *video compression*, a form of *encoding* that transforms a two-dimensional array of *pixels* (usually 8×8) to a two-dimensional array of frequency ceofficients. This transformation facilitates compression because many of the frequency coefficients have zero or small values and can be ignored or quantized with fewer bits. Statistical coding of the frequency coefficient values provides further compression. See *quantizing*.

disk drive A mechanism for recording or replaying *tracks* on a disk. It consists of a motor that rotates the disk and a record/replay head mounted on a mechanism that moves the head in a radial direction across the disk surface. Some disk drives have multiple disks and heads to increase storage capacity.

dockable In camcorders, a packaging approach that allows camera and recorder to be separated. This is useful when the camera portion is used without a recorder in a studio setting.

DTV See *digital television*.

DVB See *Digital Video Broadcasting*.

DVC See *Digital Video Cassette.*

DVD See *Digital Versatile Disc.*

dynamic range In a signal system, the ratio between the largest and the smallest signals that can be handled.

editing In *postproduction*, the process of assembling program segments cut from raw shots into a finished program sequence. Editing may also include *special effects* to provide enhanced transitions between program segments.

electronic imaging The use of electronics for storage and reproduction of natural scenes.

electronic news gathering (ENG) The use of a video camcorder for capturing news events.

electronic photography The use of electronics for capture and storage of still pictures from natural scenes.

electronic shutter In a video camera, the control of the imaging device to have an *exposure* time shorter than the normal field or frame time. This reduces blurring caused by motion in the scene and delivers better performance is slow motion or still-frame playback.

electronic zoom A zoom effect produced by selecting a reduced area of the picture and expanding it to full size by pixel and line interpolation. See *zoom lens.*

encoding The process of representing information in an electronic signal. This may be either *analog* or *digital*. Digital methods provide many more options for encoding, including *compression* techniques that reduce the required *data rate* and *bandwidth* for transmission.

ENG See *electronic news gathering.*

ergonomics The study of human-equipment interaction. It is often called *human engineering.*

exposure In photography or video, the length of time that light is integrated to capture one picture or frame. Shorter exposure times cause less blurring of moving objects in the scene but reduced *sensitivity.*

f-number A number that relates to the light-collection ability of a lens. It is calculated by dividing the *focal length* of the lens by the diameter of the lens iris aperture. Smaller f-numbers mean greater light collection.

f-stops A series of *f-numbers* that are commonly used. For example, f/2, f/2.8, f/4, f/5.6, f/8, and so on, is a progression that provides 2:1 exposure differences between stops.

FDM See *frequency-division multiplexing.*

fiber-optic cable A cable containing one or more *optical fibers* for signal transmission. It is being used more widely for digital signal transmission in communications and *camera cables*. In a camera cable, the optical fibers are often combined with two or more copper wires for transmission of power to the *camera head.*

field In *interlaced scanning*, one vertical scan period, which is the scanning of all odd lines or all even lines. Two fields are required to make a complete *frame*.

field of view In an optical system, the horizontal angle made by light rays from the left and right edges of the reproduced part of the scene. A "normal" field of view is usually considered to be about 50 degrees.

film See *photographic film*.

film transport A mechanism for moving *film* past the reading station in a *telecine system*.

film-to-tape transfer The process of reading *film* electronically in a *telecine system* and recording the output on videotape.

finite impulse response filter (FIR) A form of *digital filter* that synthesizes its output response by summing a finite number of delayed and weighted copies of the input signal.

FIR See *finite impuse response filter*.

flicker The psychophysical effect perceived by the eye when viewing an image that is periodically refreshed at too low a rate.

flyback In a scanned video display, the moving of the scanning spot from the end of one line to the start of the next line, or from the end of one *field* to the start of the next field.

flying-spot scanner A pickup system using a *CRT* to produce a scanning spot that is focused onto the image to be scanned, usually *film*. A phototube on the other side of the film picks up the transmitted light to deliver a video output.

focal length In a lens, the distance between the lens and the image plane when focused on an infinitely distant object.

foot-candle A unit of illumination equal to one *lumen* per square foot. See *lux*.

frame In a scanned video image, the complete scanning pattern including all lines and *pixels*.

frame rate The rate at which *frames* are scanned.

frame-interline transfer In a *charge-coupled device*, an architecture that combines the structures of *frame-transfer* and *interline-transfer* architectures.

frame-transfer A *charge-coupled device* architecture where the entire charge image is transferred to a storage area in a separate part of the IC chip. During active picture time, readout is from the storage area.

frame synchronizer A unit that stores one or more *frames* of *digital video* for the purpose of *synchronizing* incoming signals from remote sources to the local video system.

frequency interleaving In *composite video* systems, the choice of the color subcarrier frequency so that sideband components of the *chrominance* signal fall between the sidebands of the *luminance* signal.

frequency-division multiplexing (FDM) Multiplexing of two or more signals in the same transmission channel by having them occupy different frequency bands.

gamma In a video system, the exponent of the *gray scale* transfer characteristic.

gamma correction In a video system, nonlinear manipulation of the *gray scale* characteristic to correct for a nonlinearity elsewhere in the system. In most television system standards, gamma correction is required at the camera to correct for the inherent nonlinearity of *CRT* display devices.

geometric distortion In a video picture, spatial distortion of the image. See *barrel distortion* and *pincushion distortion*.

geometric optics The study of optical systems by determining the propagation of light rays through the system.

grating chart A test chart consisting of equally spaced horizontal and vertical lines, which may be white on a black background or black on a white background. Grating patterns are usually electronically generated and are used by themselves for testing of *registration* or with a *ball chart* for testing of camera *geometric distortion*.

gray scale In a video system, the brightness scale that is displayed. The term is often used to refer to the shape of the amplitude transfer characteristic even for signals that do not represent shades of gray.

halation A degradation of the image in an optical system caused by scattering of light within or by the optical elements. It often appears as a halo surrounding bright objects in the image.

hard disk A digital storage device, generally consisting of a rigid magnetic-coated disk written and read by *magnetic heads* configured to "fly" on an air film above the rotating disk. See *disk drive*.

HDTV See *high-definition television*.

helical scanning In videotape recording, a method to achieve a high *writing speed* by mounting *magnetic heads* to a rotating drum. The tape is wrapped around the drum so as to produce *tracks* at an angle to the edge of the tape.

high-definition television (HDTV) A new-generation television system that has up to four times the *resolution* of *SDTV* systems and a wider *aspect ratio* screen. It is a feature of *DTV* systems.

highlight In a video image, the response produced by an excessively bright spot in the scene. Highlights represent the highest video amplitudes and often have to be compressed to reduce the *dynamic range* of the signal to a displayable value.

horizontal resolution In a video system, the degree of reproduction of fine detail of vertically-oriented objects in the scene. It is quantified in *TVL*.

HSB See *hue-saturation-brightness*.

hue The attribute of color. It is the answer to the question, "What color is it?"

hue-saturation-brightness (HSB) A system of color measurement that specifies the *hue, saturation*, and *brightness* of the object.

human engineering See *ergonomics*.

IEC See *International Electrotechnical Commission*.

IEEE 1394 A standard for serial digital communication developed for video equipment. See *serial process*.

illumination The aggregate of light from all sources impinging on a scene. It is measured in *lux* or *foot-candles*.

image enhancement The process of modifying the video signals to produce a desired effect, such as greater sharpness, reduced *noise, gray scale*, color, and so forth.

image sensor See *imager*.

image stabilization a method to make the captured image of a camera immune to unintended motion of the camera, such as caused by shaking of the person carrying the camera.

imager A device for conversion of an optical image into an electrical signal.

interlaced scanning In a video system, the process of scanning half the lines in each of two vertical scans (*fields*). This is accomplished by having an odd number of total lines and making the ratio of horizontal to vertical scan frequencies equal to one-half the total line number. The result of interlacing is that large areas of the picture appear to be refreshed at the field frequency, which reduces flickering of the picture while allowing reduction of the frame rate.

interline flicker An artifact of interlaced scanning where adjacent lines containing different horizontally oriented information *flicker* at the *frame rate*.

interline-transfer In a *charge-coupled device*, an architecture where the entire charge image is transferred during vertical blanking to storage registers that are horizontally adjacent to each *pixel* of the CCD.

intermittent transport In a *telecine system*, a film transport that jumps each frame into the readout position. This allows greater exposure time for scanning the frame but it trades off transport complexity and the potential for film damage.

internal focusing In a zoom lens, all the moving parts for focusing are embedded within the lens. A focus motor is required but it has the advantage that no external part of the lens has to rotate for focusing.

International Electrotechnical Commission (IEC) An international standardizing organization.

International Standardizing Organization (ISO) An international standardizing organization.

International Telecommunications Union (ITU) An international standardizing organization. Two branches of the ITU are relevant to video: ITU-R for radio broadcasting, and ITU-T for other forms of transmission.

intracoded frames In video *compression*, when *frames* are encoded independently of each other. This allows video to be started or edited at any frame location. See *encoding, predicted frames*.

inverse DCT The process for decoding the *discrete cosine transform*.

iris In a lens, the mechanism that controls the size of the light-gathering opening.

ISO See *International Standardizing Organization*.

ITU See *International Telecommunications Union*.

Joint Photographic Experts Group (JPEG) A subcommittee of the ISO/IEC that developed an image *compression* standard for still images called JPEG compression. The standard provides a range of algorithms for compression of still images in both *lossless* and *lossy* modes.

JPEG See *Joint Photographic Experts Group*.

JPEG compression The name given to the image *compression* standard developed by the *Joint Photographic Experts Group*.

Kell factor In scanned video systems, a factor that relates the apparent *vertical resolution* (in *TV lines*) of a *raster* to the number of scanning lines in the raster. For *CRT* displays, it is generally taken as 0.7.

lag In an *imager*, a motion-smearing effect caused by incomplete readout of stored charge during a single scan. It was a serious problem with tube imagers, it is much less of a problem with *charge-coupled devices*.

Lambert's cosine law (optics) The reflected intensity from a diffuse surface varies as the cosine of the angle between the normal to the surface and the direction of viewing.

lateral chromatism In a lens, when the focal length varies with color. The result is that the images of the primary colors focus in different planes and have different magnifications, which can cause *registration* errors.

latitude In photography, the ability to obtain acceptable pictures over a range of *exposure*.

LCD See *liquid-crystal display*.

letterbox format In video displays, a method of displaying 16:9 aspect ratio pictures on a 4:3 aspect ratio screen by having black areas at the top and bottom of the picture.

limiting resolution The maximum discernable resolution of a camera or system. See *wedge chart*.

linear polarization In an electromagnetic wave, the electric field vector oscillates at a fixed angle in the plane perpendicular to the direction of propagation.

liquid-crystal display (LCD) A flat-panel display device that uses the properties of liquid crystal materials.

longitudinal chromatism See *lateral chromatism*.

look-up table In digital systems, the use of a memory to hold an array of values that are accessed by memory addressing.

lossless compression A data *compression* system that does not in any way change the information content of the data as a result of compression and decompression. Such a compressor can operate on any data without damaging it. However, the degree of compression is limited.

lossy compression A data *compression* system that may make changes to the information content of the data as a result of compression and decompression but it does that in a way that will have minimal effect on the usability of the data. This type of compression requires that the compressor know the data format and its use but, since that is known, much more compression can often be achieved than with *lossless compression*.

low-dispersion glass In a lens, a type of glass that has minimal internal scattering and transmission losses. This reduces *halation* and light loss in the lens.

lumen A unit of *luminous intensity*. One lumen is equal to the luminous flux emitted within a unit solid angle (one steradian) by a point source having a uniform intensity of 1 *candela*.

luminance In video systems, a signal that represents the visual brightness of the scene separate from its color properties.

luminosity Light flux weighted according to the *luminosity curve*.

luminosity curve A standard curve that depicts the visual brightness response to light at different wavelengths. It is used by video systems to define the *luminance* signal. It is also used in illumination science to specify light levels in terms of visual brightness.

luminous intensity The time rate of flow of light from a point source per unit solid angle in a particular direction.

lux One *lumen* per square meter. See *foot-candle*.

macro capability In a lens, the ability to focus at very short object distances.

magnetic heads In magnetic recording, the device that provides the recording magnetic field or, for replay, the device that picks up the field from the recorded surface and converts it to an electrical signal.

masking In color cameras or systems, the process of changing the color response of the system by linearly or nonlinearly mixing the three *primary color* signals.

minimum illumination In a video camera, the lowest light level needed to capture a "usable" picture. The system gain may be increased for this purpose, which trades off *SNR* for *sensitivity*.

modulation transfer function (MTF) In an imaging device or system, the curve of response to sinusoidal test patterns of steadily increasing resolution.

Moore's law In solid-state technology, an empirical statement first articulated by Gordon Moore of Intel that says: "The number of devices on an integrated circuit chip will double every 2 yr." This has proven accurate since 1965 and shows no sign of reaching an asymptote any time soon.

motion compensation In video *compression*, a technique that examines sequential *frames* and determines the parts of the scene that have changed. Only the moving parts of the scene are transmitted to construct each new frame from previous frames. This is *predicted frame* compression.

motion vector In *motion compensation*, a signal that defines how a block of pixels in a new frame can be derived from a block at a different location in a previous frame.

motion-JPEG A video *compression* method that applies the *JPEG* compression standard to each *frame* of a motion video sequence. This is an example of *intra-coded frames*.

Moving Picture Experts Group A working group of the *ISO/IEC* charged with standardization of motion video *compression* techniques.

MPEG See *Moving Picture Experts Group*.

MPEG compression A series of audio and video *compression* standards developed by the *Moving Picture Experts Group*.

MTF See *modulation transfer function*.

multiburst chart A test pattern comprising a number of groups of vertical lines of progressively closer spacing. The vertical-line groups have a sinusoidal intensity variation to support the measurement of *modulation transfer function* response of cameras and systems.

National Television Systems Committee (NTSC) The organization in the United States that developed that country's color television standard (called NTSC color television) in 1953.

ND filter See *neutral-density filter*.

neutral-density filter (ND filter) An optical filter that provides the same attenuation to light at all wavelengths. Such a filter appears gray.

NiCd See *nickel-cadmium*.

nickel-cadmium (NiCd) A type of secondary battery widely used for *camcorders*.

nickel-metal-hydride (NiMH) A type of secondary battery that provides greater capacity/weight performance than *NiCd* batteries, but at higher cost.

NiMH See *nickel-metal-hydride*.

Nipkow disk An early television camera that accomplished scanning by means of a rotating disk having a series of holes or lenses for passing light. It was developed in 1884 by Paul Nipkow of Germany.

noise In electrical systems, spurious or undesired components that are added to signals being processed or transmitted. Unavoidable random noise sources exist in most systems and place an upper limit on *signal-to-noise ratio* performance.

nonlinear editing Audio or video *editing* that uses random-access storage (usually a digital *hard disk*) to assemble edits in real time.

NTSC See *National Television Systems Committee.*

Nyquist criterion In a *sampling* system, the fact that the *sampling rate* must be at least twice the highest signal frequency to avoid *aliasing* distortion.

Nyquist limit In a *sampling* system, the highest signal frequency that can be sampled without *aliasing*. See *Nyquist criterion.*

omega wrap In a *helical-scan* videotape recorder, the situation where the tape wraps halfway or more around the scanning drum.

on-chip lens In a *charge-coupled device*, a feature where a small lens is placed over each *pixel* cell to cause all the light falling on the cell to reach its sensitive area. This is important in *interline-transfer* CCDs where the sensitive area is only a fraction of the cell's total area, which would otherwise cause loss of *sensitivity.*

optical fiber An optical transmission medium comprising a microscopic glass fiber. Light energy sent into the fiber is transmitted within the fiber by *total reflection* at the fiber's surface. Optical fibers provide wide *bandwidths* and long-distance transmission.

outside broadcast (OB) The use of video equipment for on-location *production.*

overflow drains In a *charge-coupled device*, a structure within a cell that collects excess charge formed in the cell and prevents it from spreading into adjacent cells.

PAL See *Phase-Alternating Line.*

pan and tilt head In camera mounting, a device that mounts to the top of a tripod or pedestal and allows convenient moving of the camera horizontally (panning) and vertically (tilting).

parallel process In an electronic system, where portions of the signal are processed simultaneously in separate channels.

PCMCIA An interface standard for peripheral devices in portable PCs. The acronym stands for Personal Computer Memory Card Industry Association.

persistence of vision The property of human vision to briefly retain an image flashed on the retina. This is fundamental to the working of scanned video systems and displays, as it prevents *flicker* when the frame rate is high enough.

Phase-Alternating Line (PAL) A composite analog video television standard developed in Europe and now used by many countries around the world. PAL is based on most of the concepts of *NTSC* but has some improvements. The name comes from the line-to-line alternation of the color subcarrier phase that causes some types of transmission distortions to cancel. See *composite video.*

photoconductivity The property of certain semiconductor materials to increase in conductivity when illuminated. This is used in most modern *imagers.*

photoemission The property of certain materials to emit electrons from their surface when illuminated. This is useful only in a vacuum and it is the basis of many tube-type *imagers* used in the early days of television.

photographic film A photosensitive medium made by placing one or more layers of silver halide material on a plastic substrate. Upon brief exposure to a light image, an invisible latent image is created. The latent image is made visible by subsequent chemical development that converts the latent image into actual silver molecules.

photovoltaic The property of certain materials to generate an electrical voltage when illuminated. This is used for power generation in solar cells, but is not used in imaging.

picture-in-picture A display technique where a small image of one video signal is displayed in a rectangular area of a screen that is otherwise showing a second video signal.

pincushion distortion In optical imaging, a distortion where the magnification increases away from the optical axis.

pixel In video systems, an independent point in the picture, separately definable for brightness and color. Video systems usually have a two-dimensional array of pixels; there may be hundreds of thousands or more pixels on a single *frame*.

polarization In an electromagnetic wave, the behavior of the electric field vector during propagation. See *circular polarization, linear polarization*.

polarizing filter An optical filter that provides control of the *polarization* of light waves.

polysilicon A highly conducting form of silicon used in the fabrication of ICs and *CCDs*.

postproduction In the style of program creation where all shooting (*production*) is done before editing, the process of program *editing*, assembly, and *special effects*.

potential well See *depletion region*.

predicted frames In motion compensation compression, the case where a new frame is based on the previous (or sometimes the future) frame. Because frames depend on one another, playback or *editing* cannot be started at a predicted frame.

primary colors In a color video system, the color names for the three different *taking curves* of the camera. The most common system of taking curves is *RGB*.

prism optics The light-splitting optics of most modern three-CCD cameras. Prism optics consist of solid glass prisms cemented together with appropriate *dichroic* filter layers embedded within them. This provides a highly precise and reliable light-splitting system.

production In the style of program creation where all material is captured on tape or disk before beginning program editing and assembly, the shooting and capture phase. See *postproduction*.

progressive scanning A scanning system where all lines of the *raster* are scanned in each vertical scan. See *interlaced scanning*.

quantizing In *analog-to-digital conversion*, the process of assigning a digital value to each sample.

raster In a scanned video system, the total pattern of lines produced by the *scanning motion*.

Rec. BT.601 A standard developed by ITU-R for *component* digital video. See *International Telecommunications Union*.

recording density In recording systems, the number of bits stored in a given area of the medium. For example, 1 Mb/in^2.

reflection In an optical system, the property of bouncing back the light rays at an interface between materials of different *refractive index*, as in a mirror.

refraction In an optical system, the property of bending and transmitting the light rays at an interface between materials of different *refractive index*.

refractive index The ratio of the wave propagation velocity in a medium to that in free space. In optical systems, this affects the properties of *reflection* and *refraction*.

registration In a video camera, the requirement that the separate color channels produce images that precisely overlay each other upon display.

resolution In a video system, the property of fine detail reproduction. Video resolution is measured in *TV lines of resolution*. See *horizontal resolution, limiting resolution, vertical resolution*.

return video In a video camera, the ability to display signals from other sources at the camera viewfinder.

RGB Acronym for red-green-blue, the three most common camera *primary colors*.

routing switcher In a studio system, a switching system that connects cameras, VCRs, and other units to the users who will control them.

sample and hold A *sampling* process that holds its value from one sample to the next, thus producing a stepped waveform rather than a series of pulses.

sampling The process of representing an analog signal as a series of equally spaced pulses whose amplitudes equal the value of the analog signal at their time of occurrence.

sampling rate The rate of occurrence of *sampling*.

saturation In a color representation system, the parameter that represents the intensity or depth of color. See *hue-saturation-brightness*.

scan conversion In a video system, the process of converting video data from one scanning standard to another.

scanning In a video system, the process of reading across and down an image to convert it into a time-sequential video signal. Most scanning systems use a rectangular pattern called a *raster*.

SDTV See *standard-definition television*.

SECAM See *Sequential Couleur Avec Mémoire.*

sensitivity In a video *camera*, the specification that defines the scene illumination required for full-specification performance. See also *minimum illumination.*

Sequential Couleur Avec Mémoire (SECAM) The *composite* color television system developed in France and used there and in the countries of the former Soviet Union. SECAM uses a baseband *luminance* signal and two frequency-modulated subcarriers that carry the color-difference information. See *color-difference signals.*

serial process In an electronic system, where processes occur sequentially in a single signal path.

shading In a video *camera*, picture distortion where camera black level is not uniform across the picture (black shading) or where the camera *sensitivity* varies across the picture (white shading).

shot box An automatic device for controlling a *camera* where a multiplicity of settings are remembered for a number of different scenes. Upon demand, the settings for any scene may be called up and automatically applied to the camera. A shot box may control lens settings, pan and tilt settings (when used with a servo-driven pan and tilt head), and sometimes internal camera adjustments.

signal processing In an electronic system, processes that modify, adjust, or combine signals.

signal-to-noise ratio (SNR) In an electronic system, the ratio between the normal signal level and the *noise* present with no signal or with a uniform signal. In video systems, SNR is measured in decibels and is the ratio between peak-to-peak black-to-white video and the rms noise level.

single-lens reflex (SLR) A camera whose *viewfinder* looks through the same lens that is used for capturing pictures. This can provide *WYSIWYG* performance.

SLR See *single-lens reflex.*

SMPTE See *Society of Motion Picture and Television Engineers.*

SNR See *signal-to-noise ratio.*

Society of Motion Picture and Television Engineers (SMPTE) An organization of engineers that develops standards in the fields of audio, video, recording, and motion pictures.

spatial compression Video *compression* techniques that exploit the redundancy between *pixels* and lines to accomplish compression.

special effects In *postproduction*, the use of dynamic transitions between video signals, such as wipes, dissolves, rotations, page-turnings, and so forth.

specularity In a natural scene, an area that reflects the illumination source directly at the camera lens. This produces an extreme highlight that has the same color as the illumination, for example, a reflective round ball will have a specularity on its surface at the location that satisfies the above condition.

spherical aberration In a lens, an *aberration* caused by light rays that pass through the edges of a lens focusing at a different distance than rays passing near the axis of the lens.

stairstep chart A test pattern displaying a series of gray areas of increasing reflectance. This is used for testing camera or system *gray scale* response.

standard-definition television (SDTV) Television systems having the same scanning standards as the existing analog television systems, such as *NTSC*, *PAL*, or *SECAM*. The term is used loosely and sometimes also refers to modifications of the existing standards, such as using progressive scanning or 16:9 aspect ratio with 525 lines, or using the 640 × 480 pixel computer standard for television.

still-picture camera A camera (photographic or video) that captures single *frames*.

subtractive color A color reproduction system where display is accomplished by placing colored dyes on a white reflective surface. Uses of this are in color printing or color print photography. The primaries for subtractive color are usually cyan, magenta, and yellow, although a fourth color (black) may also be added to improve black rendition, especially in color printing. That system is called CMYK.

symbol In a digital transmission system, the basic unit of data carried at one moment by the channel signal. A single symbol may carry 1 or more bits.

sync generator A circuit that generates the horizontal and vertical timing signals to control the scanning of multiple video devices so they will be synchronous.

taking curves In a color video camera, the optical spectral responses of the three color channels.

tape transport In a recording system, the mechanism that handles unrolling of tape from the supply reel, passing it over the *magnetic heads* or the *helical scanner*, and rolling it up on a takeup reel. Other tape transport functions may include cassette loading and tape rewinding.

TDM See *time-division multiplexing*.

telecine system A special-purpose video camera system designed for video reproduction from motion picture film.

telephoto A lens that has a *field of view* smaller than the 50-degree "normal" lens. Thus, it has a longer than normal *focal length*.

teleprompter A video display device mounted on a camera to display prompt messages to performers in front of the camera. Usually, the camera looks through a half-silvered mirror on the teleprompter so that messages are displayed directly over the camera lens and can be read by performers when looking directly at the camera.

television A video system specifically designed for distribution of pictures and sound to a mass audience. Distribution may be by means of broadcasting, either terrestrial or satellite, or by cable.

temporal compression Video *compression* techniques that exploit the redundancy between successive *frames* of a video sequence. See *motion compensation*.

thick lens A lens whose thickness may not be neglected compared to the focusing distances. This is usually the case with lenses having multiple optical elements.

thin lens A lens whose thickness can be neglected compared to the focusing distances.

time code In recording systems for audio or video, a special code recorded on the same medium along with the audio or video information for the purpose of uniquely identifying video frames and the corresponding audio. This is necessary for precise *editing* of audio and video.

time-division multiplexing (TDM) In an electronic system, the combining of multiple signals in the same channel by assigning them segments of time. TDM is commonly used for digital multiplexing.

total reflection In an optical reflecting situation, when the angle of incidence with a medium that has a lower *refractive index* is less than a critical value. At this condition and below, there is no *refraction* and all incident light is reflected.

tracks The pattern recorded on a recording medium to provide a single recording channel. Tracks on tape may be longitudinal (along the tape) or *helical* (at an angle to the tape edge). Tracks on disk are either concentric circles or spirals.

transfer smear In a *charge-coupled device*, a distortion where highlights in the scene interfere with the charge transfer process. It shows up as a vertical trail above and below highlights in the picture.

transparent The property of a system to not change signals passing through it.

triax cable A three-conductor *coaxial cable* often used for *camera cables*. Triax consists of a center conductor and two coaxial shield layers. The center conductor and the inner shield function as normal coaxial cable for signal transmission; the outer shield provides additional protection against interference needed with long cable lengths.

trichromatic color A color reproduction system having three independent channels with different spectral responses to light. The three spectral responses are referred to by their dominant colors, such as red, green, and blue; these are the *primary colors* for the system.

TV See *television*.

TV lines of resolution (TVL) The measure of *resolution* in a video system. TVL is defined as the number of equally-spaced white and black lines in a distance equal to the picture height.

TVL See *TV lines of resolution*.

vertical resolution Picture *resolution* in the vertical direction; that is, observed on horizontally oriented lines or patterns.

video The electronic representation of pictures, either stationary or moving.

video compression The application of data *compression* technology to video signals.

video recorder A recording device designed to record video signals.

video system A system for capturing, recording, processing, or reproduction of video.

viewfinder In a camera, the device that allows the camera operator to see the area of the scene that is being captured. Viewfinders may be optical or electronic.

viewing ratio In viewing a video display, the ratio between the distance from viewer to display and the picture height. The closer the viewer sits to a given display, the smaller the viewing ratio and the greater *resolution* he or she will be able to perceive on the display.

vignetting In an optical system or imager, the tendency for sensitivity to fall off toward the edges of the image, causing a "porthole" effect in the reproduced picture.

visible light Light in the wavelength range from 400 to 700 nm. This is the spectral range of normal human vision.

waveform monitor A display device for observing video or other waveforms. It is a special form of oscilloscope that is adapted to have built-in capabilities for synchronizing to video signals.

wavelength In any wave phenomenon, the ratio between the velocity of propagation and the wave frequency.

wedge chart For measurement of *limiting resolution*, a chart consisting of a pattern of white and black lines that converge, thus requiring higher and higher resolution for successful reproduction. The pattern may be calibrated in *TVL*; the farthest point down the chart where the lines are still visible would be the system's limiting resolution. Wedge charts can be displayed at any angle in the picture to test different angular resolutions.

white balance In a video system, the condition where perceived whites in the scene are reproduced as the specified white condition for the system. For example, most *RGB* systems specify white as equal R, G, and B signals.

wide-angle A lens whose *field of view* is wider than the "normal" 50 degrees.

writing speed In a recording system, the speed at which tracks pass by the recording or replaying head.

WYSIWYG A computer industry acronym that means "what you see is what you get." It is applicable to cameras whose *viewfinder* provides an exact replica of the picture being captured.

YC_RC_B format A component video format consisting of *luminance* (Y) and two color-difference components based on $R - Y$ and $B - Y$. See *color-difference signals*.

zebra processing In a camera *viewfinder*, a technique that displays one or more patterns in the display at locations where the video level exceeds specified values. It allows the operator to manually set exposure without viewing a video waveform.

zoom lens A lens of variable *focal length*.

Bibliography

GENERAL VIDEO AND TELEVISION

Baron, S. N. (ed.), *Implementing HDTV: Television and Film Applications*, SMPTE, White Plains, NY, 1996.

Benson, K. B., and Whitaker, J., *Television Engineering Handbook*, McGraw-Hill, New York, 1992.

Fisher, D. E., and Fisher, M. J., *TUBE: The Invention of Television*, Counterpoint, Washington, DC, 1996.

Inglis, A. F., and Luther, A. C., *Video Engineering, 2nd ed.*, McGraw-Hill, New York, 1996.

DIGITAL AND COMPUTER TECHNOLOGY

Dorf, R. C., *Electrical Engineering Handbook*, CRC Press, Boca Raton, FL, 1993.

Luther, A. C., *Principles of Digital Audio and Video*, Artech House, Norwood, MA, 1997.

Jayant, N. S., and Noll, P., *Digital Coding of Waveforms*, Prentice Hall, Englewood Cliffs, NJ, 1984.

Negroponte, N., *Being Digital*, Knopf, New York, 1995.

Rzeszewski, T. S. (ed.), *Digital Video: Concepts and Applications Across Industries*, IEEE Press, New York, 1995.

Watkinson, J., *The Art of Digital Video, 2nd ed.*, Focal Press, London, 1994.

TELECOMMUNICATIONS

Agnew, P. W., and Kellerman, A. S., *Distributed Multimedia: Technologies, Applications, and Opportunities in the Digital Information Industry*, ACM Press, New York, 1996.

Gibson, J. D., (ed.), *The Communications Handbook*, CRC Press, Boca Raton, FL, 1997.

Inglis, A. F., and Luther, A. C., *Satellite Technology: An Introduction, 2nd ed.*, Focal Press, Boston, 1997.

Keiser, *Broadband Coding, Modulation, and Transmission Engineering*, Prentice Hall, Englewood Cliffs, NJ, 1989.

Lu, G., *Communication and Computing for Distributed Multimedia Systems*, Artech House, Norwood, MA, 1996.

Minoli, D., and Keinath, R., *Distributed Multimedia Through Broadband Communications Services*, Artech House, Norwood, MA, 1994.

Ohta, N., *Packet Video: Modeling and Signal Processing*, Artech House, Norwood, MA, 1994.

Riley, M. J., and Richardson, I. E. G., *Digital Video Communications*, Artech House, Norwood, MA, 1997.

Schaphorst, R., *Videoconferencing and Videotelephony: Technology and Standards*, Artech House, Norwood, MA, 1996.

OPTICS

Canon, Inc., *TV Optics II, The Canon Guide Book of Optics for Television Systems*, 1992.

Meyer-Arendt, J. R., *Introduction to Classical and Modern Optics, 4th ed.*, Prentice-Hall, Englewood Cliffs, 1995.

AUDIO

Benson, K. B., *Audio Engineering Handbook*, McGraw-Hill, New York, 1988.

Pohlmann, K. C., *Principles of Digital Audio, 3rd ed.*, McGraw-Hill, New York, 1995.

VIDEO CAMERAS AND POSTPRODUCTION

Anderson, G., *Video Editing and Postproduction: A Professional Guide, 3E*, Focal Press, Boston, 1993.

Zettl, H., *Television Production handbook, Sixth Edition*, Wadsworth Publishing, Belmont CA, 1997.

STANDARDS

Advanced Television Systems Committee (ATSC), http://www.atsc.org.

Digital Video Broadcasting (DVB), http://www.dvb.org.

Institute of Electrical and Electronics Engineers (IEEE), http://www.stdsbbs.ieee.org.

International Standards Organization (ISO), http://www.iso.ch.

International Telecommunications Union (ITU), http://www.itu.ch.

Society of Motion Picture and Television Engineers (SMPTE), White Plains, NY. http://www.smpte.org.

PERIODICALS

AV Video Magazine, published monthly by Montage Publishing, Inc., 701 Westchester Ave., White Plains, NY 10604.

Byte Magazine, published monthly by the McGraw-Hill Companies, Inc., P.O. Box 552, Hightstown, NJ 08520.

IEEE Transactions on Consumer Electronics, IEEE Operations Center, 445 Hoes Lane, P. O. Box 1331, Piscataway, NJ 08855-1331.

IEEE Transactions on Broadcasting, IEEE Operations Center, 445 Hoes Lane, P. O. Box 1331, Piscataway, NJ 08855-1331.

Keyboard Magazine, published monthly by Miller-Freeman, Inc., 600 Harrison St., San Francisco, CA 94107. http://www.keyboardmag.com.

Millimeter Magazine, published monthly by Intertec Publishing, Corporation, 5 Penn Plaza, 13th Floor, New York, NY 10001.

Popular Photography Magazine, published monthly by Hachette Filipacchi Magazines, Inc., P.O. Box 54912, Boulder, CO 80322-4912.

SMPTE Journal, published monthly by the Society of Motion Picture and Television Engineers, White Plains, NY. http://www.smpte.org.

Videography Magazine, published monthly by Miller-Freeman PSN, Inc., 460 Park Avenue South, New York, NY 10016.

Videomaker Magazine, published monthly by Videomaker, Inc., P.O. Box 4591, Chico, CA 95927.

OTHER WEB SITES

http://www.alegria.comtelecinehome.html (telecine).

http://www.amazon.com (bookseller).

http://www.antonbauer.com (batteries).

http://www.belden.com (wire and cable).

http://www.canon.com (cameras, optics).

http://www.hitachi.com (cameras, camcorders).

http://www.jvc-america.com (cameras).

http://www.kodak.com (digital still cameras, film and telecine).

http://www.oconnor.com (camera mounting equipment).

http://www.olympusamerica.com (digital still cameras).

http://www.panasonic.com (cameras and camcorders).

http://www.philipsbts.com (cameras, telecine).

http://www.ricohcpg.com (digital still cameras).

http://www.sarnoff.com (video research).

http://www.sharp-usa.com (camcorders).

http://www.sel.sony.com (cameras and camcorders).

http://www.snellwilcox.com (standards conversion, synchronizers).

http://www.tek.com (measurement).

http://www.vinten.com (camera mounting equipment).

Index

Symbols

3:2 pulldown   153
4:2:0   264
4:4:4   21

A

aberrations   33
ACATS. *See* Advisory Committee on
 Advanced Television Systems
active scan time   11
ADC. *See* analog-to-digital conversion
add-on editing   125
additive color   6
Advanced Television Systems Committee   97
Advisory Committee on Advanced
 Television Systems   97
algorithm   92
aliasing   9, 60
altitude   206
analog-to-digital conversion   8
antialiasing   260
aperture correction   75
application-specific IC   270
architectures   53
ASIC. *See* application-specific IC
aspect ratio   11
aspherical lenses   37, 272
ATSC. *See* Advanced Television
 Systems Committee
automatic gain control   81

B

batteries   216

nickel-cadmium   216
 nickel-metal-hydride   216
Betacam SX   210
black burst signal   117
black shading   88
blanking intervals   11
brightness   5, 16

C

camcorders   169
 dockable   183
 hand-held   180
 shoulder-held   180
camera cable   132
 fiber-optic   138, 262
 multiconductor   132
 triax   135
camera control unit   119
camera signal processing   75
cameras
 camcorders   169
 future   269
 hand-held   180
 HDTV   257
 measurements   243
 sensitivity   240
 shoulder-held   180
 signal processing   75
 specifications   239
 still picture   221
candela   26
CCD. *See* charge-coupled devices
CCU. *See* camera control unit
characteristic impedance   134

The Artech House Audiovisual Library

Communication and Computing for Distributed Multimedia Systems, Guojun Lu

Computer-Mediated Communications: Multimedia Applications, Rob Walters

Digital Video Communications, Martyn J. Riley and Iain E. G. Richardson

Distributed Multimedia through Broadband Communications Services, Daniel Minoli and Robert Keinath

Fax: Facsimile Technology and Applications Handbook, Kenneth McConnell, Dennis Bodson, and Richard Schaphorst

Networks and Imaging Systems in a Windowed Environment, Marc R. D'Alleyrand, editor

Packet Video: Modeling and Signal Processing, Naohisa Ohta

Principles of Digital Audio and Video, Arch C. Luther

Television Technology: Fundamentals and Future Prospects, A. Michael Noll

Video Camera Technology, Arch C. Luther

Videoconferencing and Videotelephony: Technology and Standards, Richard Schaphorst

For further information on these and other Artech House titles, including previously considered out-of-print books now available through our In-Print-Forever™ (IPF™) program, contact:

Artech House
685 Canton Street
Norwood, MA 02062
781-769-9750
Fax: 781-769-6334
Telex: 951-659
e-mail: artech@artech-house.com

Artech House
Portland House - Stag Place
London SW1E 5XA England
+44 (0) 171-973-8077
Fax: +44 (0) 171-630-0166
Telex: 951-659
e-mail: artech-uk@artech-house.com

Find us on the World Wide Web at:
www.artech-house.com